Ecological Studies

Analysis and Synthesis

Edited by

W. D. Billings, Durham (USA) F. Golley, Athens (USA)

O. L. Lange, Würzburg (FRG) J. S. Olson, Oak Ridge (USA)

Volume 24

J. N. Kremer · S. W. Nixon

A Coastal Marine Ecosystem

Simulation and Analysis

With 80 Figures

Springer-Verlag Berlin Heidelberg New York 1978

Professor Dr. JAMES N. KREMER
Department of Biological Sciences
University of Southern California
Los Angeles, CA 90007/USA

Professor Dr. SCOTT W. NIXON
Graduate School of Oceanography
University of Rhode Island
Kingston, RI 02881/USA

ISBN-13: 978-3-642-66719-0 e-ISBN-13: 978-3-642-66717-6
DOI: 10.1007/978-3-642-66717-6

Library of Congress Cataloging in Publication Data. Kremer, James N. 1945— . A coastal marine ecosystem. (Ecological studies; v. 24). Bibliography: p. Includes index. 1. Marine ecology—Mathematical models. 2. Marine ecology—Data processing. 3. Coastal ecology—Mathematical models. 4. Coastal ecology—Data processing. I. Nixon, Scott W., 1943— joint author. II. Title. III. Series. QH541.5.S3K73. 574.5'26. 77-22785.

Softcover reprint of the hardcover 1st edition 1978

Preface

> One aim of the physical sciences has been to give an exact picture of the material world. One achievement of physics in the twentieth century has been to prove that that aim is unattainable.... There is no absolute knowledge. And those who claim it, whether they are scientists or dogmatists, open the door to tragedy. All information is imperfect. We have to treat it with humility.
>
> Bronowski (1973)
> The Ascent of Man

It seems particularly appropriate to us to begin this book with Jacob Bronowski's passionate message firmly in mind. Those who set out to construct numerical models, especially ones that are mechanistic and essentially deterministic, must work always with this awareness as a backdrop for their efforts. But this is also true for the most meticulous physiologist or observant naturalist. We are all dealing with simplifications and abstractions, all trying to figure out how nature works. Unfortunately, this common pursuit does not always lead to mutual understanding, and we have become increasingly aware over the past six years that many ecologists feel a certain hostility or at least distrust toward numerical modeling. In a number of cases the reasons for such feelings are personal and very understandable—hard-gotten data skimmed off by someone with little appreciation for the difficulties involved in obtaining reliable measurements, grandiose claims of predictability, the tendency for some model builders to treat other scientists as number-getters whose research can be directed in response to the needs of the model, etc. In other cases, however, it is often a lack of understanding of what is actually involved in building a mechanistic numerical model and what its uses and limitations are. Published articles tend to describe models briefly and in strictly mathematical terms, so as to stress some particular applications. Moreover, much of the modeling literature is very theoretical and appears to be concerned more with the behavior of differential equations than with living systems. It is difficult for the practicing biologist or ecologist who measures metabolic rates, or nutrient kinetics, or feeding patterns, to relate his efforts to equations with a few highly aggregated coefficients.

Part of our hope in writing this book is to make some contribution toward a synthesis of effort among those who work in the laboratory and the field, and those who are dealing with nature on a more abstract level. Or, if not a synthesis of effort, at least an increase in tolerance. For this reason, a large portion of the book is devoted to a review and discussion of the empirical basis for the choices of various functional forms and the values of coefficients. The difficulties and uncertainties involved in arriving at many values serve to reinforce the caution of those who make the measurement, but also to emphasize the value of numerical systems analysis in exploring the implications and consequences of that uncertainty.

The Narragansett Bay model is concerned largely with a mechanistic numerical description of phytoplankton-zooplankton-nutrient dynamics in a temperate estuary, and the ways in which these dynamics are influenced by light, temperature, and hydrography. While the effects of some higher trophic levels are also part of the model, there is virtually no mechanistic detail included for the microbial aspects of the system. Bacterial decomposition and regeneration are simulated entirely through empirical regressions relating net rates to temperature. This crude representation is primarily a consequence of the general lack of detailed knowledge about microbial dynamics in coastal marine ecosystems. Fortunately, this situation is changing rapidly, and many of the recent advances in marine microbial ecology will come to influence future ecosystem modeling efforts. In fact, the next volume in the Ecological Studies series edited by Dr. G. Rheinheimer presents an extensive discussion of the microbial ecology of Kiel Bight, with emphasis on the role of marine bacteria and fungi.

For the most part, the mathematics required in mechanistic ecosystem modeling are not elaborate. A few basic forms appear repeatedly, and an introductory calculus course should make the reader reasonably comfortable with the expressions used here. In order to serve as a review of some principles that are particularly appropriate to understanding the model, however, we have included a chapter on the mathematical concepts involved.

As anyone who has worked with simulation models is painfully aware, the development of the computer program or algorithm for the model is an integral part of the effort. It is no trivial undertaking to develop an efficient program for a system as complex as the Narragansett Bay model. For the most part, the translation of system-flow diagrams and equations into a computer program is discussed neither in descriptions of various ecosystem models nor in programming manuals. It has become increasingly clear to us that many people who are not computer programmers are completely mystified by this step in the modeling process. In some cases, the programming is dismissed as if it involved nothing more intellectually challenging than key punching. There are, of course, many levels on which one can discuss the development of simulation algorithms, and this book does not attempt to teach computer programming. However, to give some feeling for the problem involved in developing the Narragansett Bay model, we have included a discussion of the algorithm with examples at various levels of programming complexity. Even those with no knowledge of computer languages should be able to gain some appreciation of the process and the effort involved. For the experienced programmer, the model provides some intriguing challenges. We have described some of our solutions, but there are surely others equally as good or better.

Much of the preceding has described the effort we have made to provide a description of a numerical ecosystem model that will be useful to those who do not really consider themselves systems ecologists or modelers. However, we have also tried to provide a rigorous and detailed description of our particular model that will be of interest to others in this field. While it was developed for Narragansett Bay, there is really nothing in the model except for some of the forcing functions and the choices of initial conditions that ties it to this particular area. In the more general sense, it is a model of a temperate, plankton-based marine ecosystem with relatively constant salinity and a well-mixed water column. Our aim in developing the model has not been to get the best fit to a set of observed points or to make direct predictions about what the bay might do in the future. We have used the model to synthesize a great deal of information on marine ecosystems in general, and on Narragansett Bay in particular. We have then tried to go the other way and to use the model in an analysis of the bay ecosystem, to explore a number of hypotheses about which processes may or may not be important in forming the patterns that emerge in the natural system. We have tried not to confuse numbers with knowledge.

Autumn 1977

JAMES N. KREMER
SCOTT W. NIXON

Contents

1. Perspectives

There is an underlying assumption in much of ecology, and perhaps science in general, that one learns about nature by isolating successively smaller parts for detailed measurement and study. This reductionism is implicit in a vast and growing literature devoted to the description of rates and responses of a variety of organisms under controlled laboratory conditions. While some of this work is done for the purpose of studying physiological processes per se, a great deal of it is undertaken in the hope that the results will reveal something about the role of the organism or species in nature. However, it is extremely difficult to perceive intuitively the consequences of any set of values in a complex, dynamic ecosystem with subtle feedback controls. One approach to this problem has led to the development of numerical simulation models, which can serve as powerful synthetic tools with which to assess quantitatively the consequences and consistency of complex sets of hypotheses. Such models may differ widely with respect to the processes and environments they are designed to represent, but in general all follow a similar developmental process and are all based on the same historical conception.

1.1 The Evolution of Ecosystem Models

The utility of numerical models in the study of marine ecosystems has been recognized for at least 35 years. In one of the earliest attempts to formulate aspects of the biological dynamics of the sea in mathematical terms, Fleming (1939) examined seasonal changes in phytoplankton population levels in the English Channel using a simple differential equation:

$$\frac{dP}{dt} = P[a-(b+ct)]\,.$$

Thus the rate of change with time of the phytoplankton population depended upon the biomass (P), a constant division rate (a), and a grazing rate composed of an initial value (b) and an arbitrary rate of increase ($c \cdot t$).

Except for the additional term, 'c', Fleming's description was essentially identical to the basic exponential growth formulation given earlier by Lotka (1925), Volterra (1926), and others, for the change in a single population (N) as a function of a constant birth and death rate.

$$\frac{dN}{dt} = N(b-d).$$

Fleming's treatment was modified somewhat by Riley and Bumpus (1946), who applied it to an analysis of phytoplankton growth on Georges Bank. As a logical improvement, they introduced another term, a_1, which represented the rate of change in phytoplankton division rate:

$$\frac{dP}{dt} = P[a + a_1 t - (b + ct)].$$

In subsequent papers, Riley (1946, 1947a) expanded the basic equation to include a number of mechanistic processes thought to be important in regulating phytoplankton populations on Georges Bank and in New England coastal waters. He obtained impressive agreement between calculated and observed populations using the formulation:

$$\frac{dP}{dt} = P[(p \cdot \bar{I})(1-N)(1-V) - R_0 \mathrm{e}^{rt} - \mathrm{g}Z].$$

The net rate (in brackets) includes a growth estimate balanced against losses due to respiration as an exponential function of temperature, as well as to zooplankton grazing. The growth estimate, in turn, includes a maximum as a linear function of the average light throughout the euphotic zone, which is reduced by two factors representing nutrient depletion of phosphorus and vertical turbulence.

As part of his study of plankton dynamics on Georges Bank, Riley (1947a) also presented a preliminary formulation in this same manner for herbivorous zooplankton:

$$\frac{dH}{dt} = H(A - R - C - D).$$

Here, too, increased biological detail was included, projecting the change in herbivores per unit biomass as the sum of a constant assimilation balanced against temperature-dependent respiration, carnivorous predation by chaetognaths, and a statistically estimated additional natural mortality. It should be pointed out that Riley projected the population changes using a mathematically exact method for successive two-week time intervals, rather than by the approximation methods used with many present digital computer models.

It was not until two years later that Riley et al. (1949) coupled the mechanistic phytoplankton and zooplankton equations together in a feedback system that was used to calculate the steady-state plankton population levels for various areas of the Western North Atlantic. While such a synthesis was conceptually straightforward, it represented an impressive mathematical and computational achievement in the days before high-speed digital computers. Nevertheless, the lack of time-varying solutions to the coupled equations severely limited their application in marine ecology for more than ten years after the initial advances of Riley and his coworkers. The full development of dynamic, mechanistic feedback models in ecology did not really begin until the application of analog and digital computers in the 1960s (Wiegert, 1975). Throughout the 1950s, the few attempts to pursue this type of model reverted to more detailed analyses of primary production (Ryther and Yentsch, 1957; Steele, 1958; Cushing, 1959) or grazing (Cushing, 1968; Harris, 1968).

It is of historical interest that another group of marine ecologists working primarily in fisheries management joined with theoretical population ecologists from many backgrounds in using the much simpler Lotka-Volterra equations to study predator-prey-interactions.

$$\frac{dN_1}{dt} = N_1(b_1 - d_1 N_2) \quad \text{prey}$$

$$\frac{dN_2}{dt} = N_2(b_2 N_1 - d_2) \quad \text{predator}.$$

Even in coupled form, it was possible to obtain dynamic solutions to the population growth equations by hand calculations, if the coefficients remained simple.

Thus, before the analog and digital computers became widely available, there was a trade-off in ecological modeling. If one wanted to explore the behavior of one or two populations over time, then it was necessary to use simple equations with very general coefficients. However, if one wanted to use a model to do work in ecosystem analysis, it was critical to have coefficients with significant biological meaning and detail. Yet without the computer, such detailed equations could only be solved for steady-state conditions under various sets of assumptions. These different goals, methods of approach, and constraints produced a divergence between ecologists interested in ecosystem analysis and those interested in population dynamics. The increasing use of computers may eventually bring the two groups together, since the development of numerical techniques has made it possible to have an almost unlimited amount of mechanistic detail with time-varying solutions. For example, one of the active areas of investigation in fisheries modeling is the introduction of time-delay terms in population growth to more accurately describe recruitment (Marchessault, 1974). While this additon introduces nowhere near the complexity that has increasingly characterized models of lower trophic levels, it is following the same trend.

With the increasing availability of analog and digital computers in the 1960s, the removal of the steady-state constraint in mechanistic models stimulated renewed interest in their potential application to a great variety of ecological problems. As mentioned earlier, essentially all of these models are extensions of the historical trend toward increasing mechanistic detail. Of course, each particular model has its own variations. The literature on this subject is growing rapidly, and has been repeatedly discussed elsewhere. Historical reviews of mechanistic ecological models have been prepared by Riley (1963), Patten (1968), Margalef (1973), Steele (1974), Wiegert (1975), and others. An introduction to the principles of ecological modeling with examples of various types of models is available in a series of volumes edited by Patten (1971–1975) and in a general text edited by Hall and Day (1977). The proceedings of several recent conferences have emphasized the use of models in understanding estuarine (Cronin, 1975) and marine (Nihoul, 1975) ecosystems, as well as the role of models in ecosystem analysis and prediction (Levin, 1975). In the marine environment, most ecological models have been developed for pelagic systems such as the North Sea (Steele, 1958), the Continental Shelf off Florida (O'Brien and Wroblewski, 1972), and upwelling regions (Walsh and Dugdale, 1971; Walsh, 1975), though models have also been derived for kelp beds (North, 1967), salt marshes (Pomeroy et al., 1972; Nixon and Oviatt, 1973a), and seagrass beds (Short, in press). A recent volume of The Sea has been devoted to discussions of the numerical modeling of physical, geological, chemical and biological processes in estuarine and open ocean waters (Goldberg et al., 1977).

1.2 The Modeling Process

A mechanistic numerical model begins with observations of the "real world". From these observations emerge tentative answers to questions about the system which may be appropriate to a modeling analysis. What is our concept of the system? What are its physical and temporal boundaries? What are the major compartments in it and how do they vary in space and time? What flows of matter, energy, or information connect these compartments? What are the important forcing functions or inputs to and outputs from the system? What time scales are involved in the major processes of interest?

Over 20 years of laboratory and field work have made Narragansett Bay one of the most intensively studied and well-known marine ecosystems. As a result, we were able to draw on the data and experience of many scientists with specialized knowledge of various aspects of the system. A number of additional studies were carried out directly as part of the modeling effort, including a year-long sampling program of phytoplankton, zooplankton and nutrients at 13 stations around the bay. These data were collected specifically for use in verifying the model. The internal formulations of the model were based on general ecological and physiological principles combined with separate historical data and independent laboratory measurements.

Various techniques were used to organize this information, including verbal summaries, tables, graphs, budgets, and flow diagrams. The resulting conceptual model was much simpler than nature and yet far too complicated to simulate. For example, even a moderately detailed energy-flow diagram for Narragansett Bay on a summer day (see Fig. 7) contains detail that would make a mechanistic model extremely cumbersome if all compartments were simulated. The decision of how much detail to include in a model is always extremely difficult. A model by definition is a simplification, and much of its utility is due to the fact that it lacks the bewildering complexity of the "real world". Yet there must also be sufficient detail to give the model credibility, to make it useful as a tool in synthesizing a variety of measurements, and to provide the parameters for a revealing sensitivity analysis of the system.

In the Narragansett Bay model, the emphasis was placed on representing the major features of the plankton-nutrient components of the system. The generalizations in the conceptual and mathematical formulations that were made toward this goal will be discussed in detail in the following sections. However, before becoming enmeshed in the details and differential equations that are the structure and function of the model, a brief description of the natural system may be useful.

1.3 Narragansett Bay

1.3.1 Physical Setting

Narragansett Bay runs roughly north–south in the Rhode Island coast between Long Island, N.Y., and Cape Cod, Mass. (Fig. 1). Immediately offshore are the waters of Rhode Island Sound and Block Island Sound. A drainage basin of some

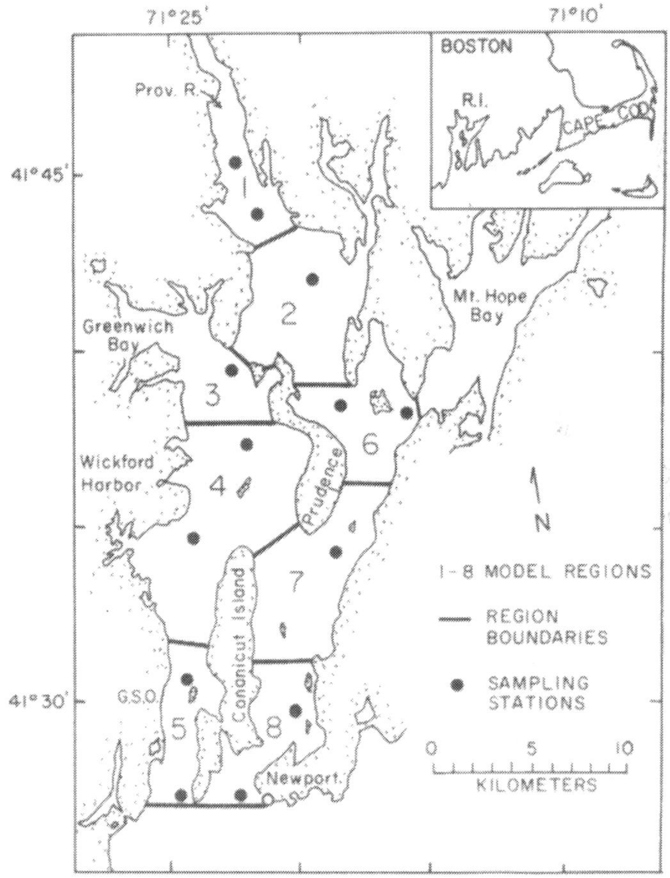

Fig. 1. Narragansett Bay, Rhode Island, and its location on the New England coastline. *Heavy dots*: stations sampled over an annual cycle to collect zooplankton, phytoplankton, and nutrient data for comparison with model simulations. The eight spatial elements or ecological subsystems of the bay were coupled by a hydrodynamic mixing model

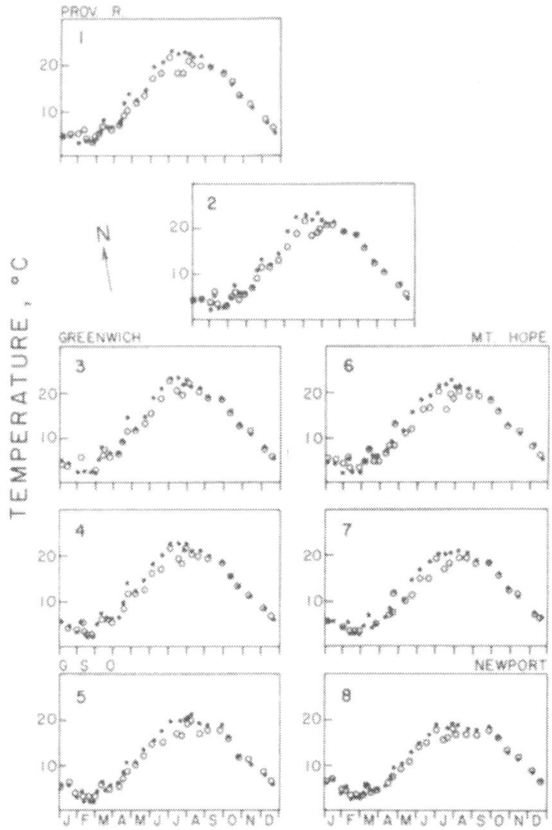

Fig. 2. Surface (●) and bottom (○) water temperatures in Narragansett Bay from Aug. 1972–Aug. 1973. Measurements were taken at 13 stations around the bay and averaged for each of the eight spatial elements of the ecosystem model (see Fig. 1)

4660 km^2 (1800 mi^2) in Rhode Island and Massachusetts provides fresh water input varying seasonally around an average of 37 m^3/s (1300 cfs). The general oceanography of the bay has been described by Hicks (1959) and in a series of reports edited by Fish (1953). The geological development and sediment characteristics have been discussed and studied by Shaler et al. (1899) and by McMaster (1960). In general, silt-clay sediments dominate the upper bay, with fine sands near the mouth. Suspended sediment loads range from one to five mg/l (Morton, 1972; Oviatt and Nixon, 1975). The bay model encompasses an area of 265 km^2 with a length of 45 km and a maximum width of 18 km. Mean depth in the bay is about 9 m, with averages of 7.5 m in the West Passage and 15.2 m in the East Passage. A hypsographic curve for the model area shows that 75% of the bay is shallower than 12 m. Tides are semidiurnal with a mean range of 1.1 m at the mouth and 1.4 m at the head. The mean tidal prism is about 13% of the mean volume and over 250 times the mean total river flow during a tidal cycle. In addition to large energy inputs associated with the tides (Levine, 1972), the waters of the bay are well-mixed by

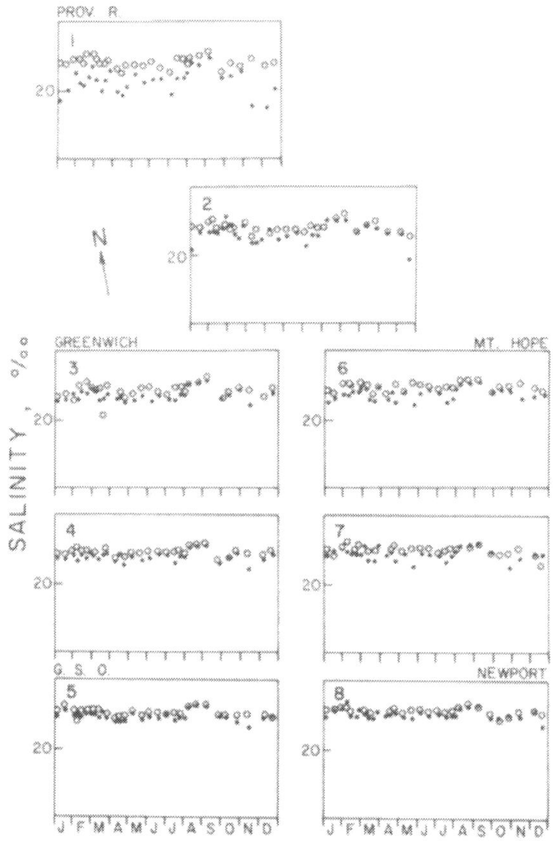

Fig. 3. Surface (●) and bottom (○) salinity in Narragansett Bay from Aug. 1972–Aug. 1973. Measurements were taken at 13 stations around the bay and averaged for each of the eight spatial elements of the ecosystem model (see Fig. 1)

winds that may completely dominate the short-term circulation pattern in some sections of the bay (Weisberg and Sturges, 1973). The wind pattern shifts markedly from northwest in the winter to southwest during summer with greatest speed during December and January (Nixon and Kremer, 1977). Water temperatures range from about $-0.5°$ C to 24° C with a well-developed thermocline present only in the upper bay and river during summer (Fig. 2). During extreme conditions, the temperature range of the surface water within the bay may be on the order of 10° C. The annual temperature cycle lags solar radiation by about 40 days. Rainfall is about 1 m year^{-1}, with river discharge usually showing a marked peak in March and April. Detailed measurements of light extinction coefficients throughout the bay have been made by Schenck and Davis (1972) and show clear gradients that reflect distance from the river and the tidal circulation pattern. Values range from about 0.20 m^{-1} in the lower East Passage to almost 1.0 m^{-1} in the river.

The relatively small fresh-water input and large tidal volume of the bay result in a well-mixed water column and small salinity gradients down the bay (Fig. 3). There

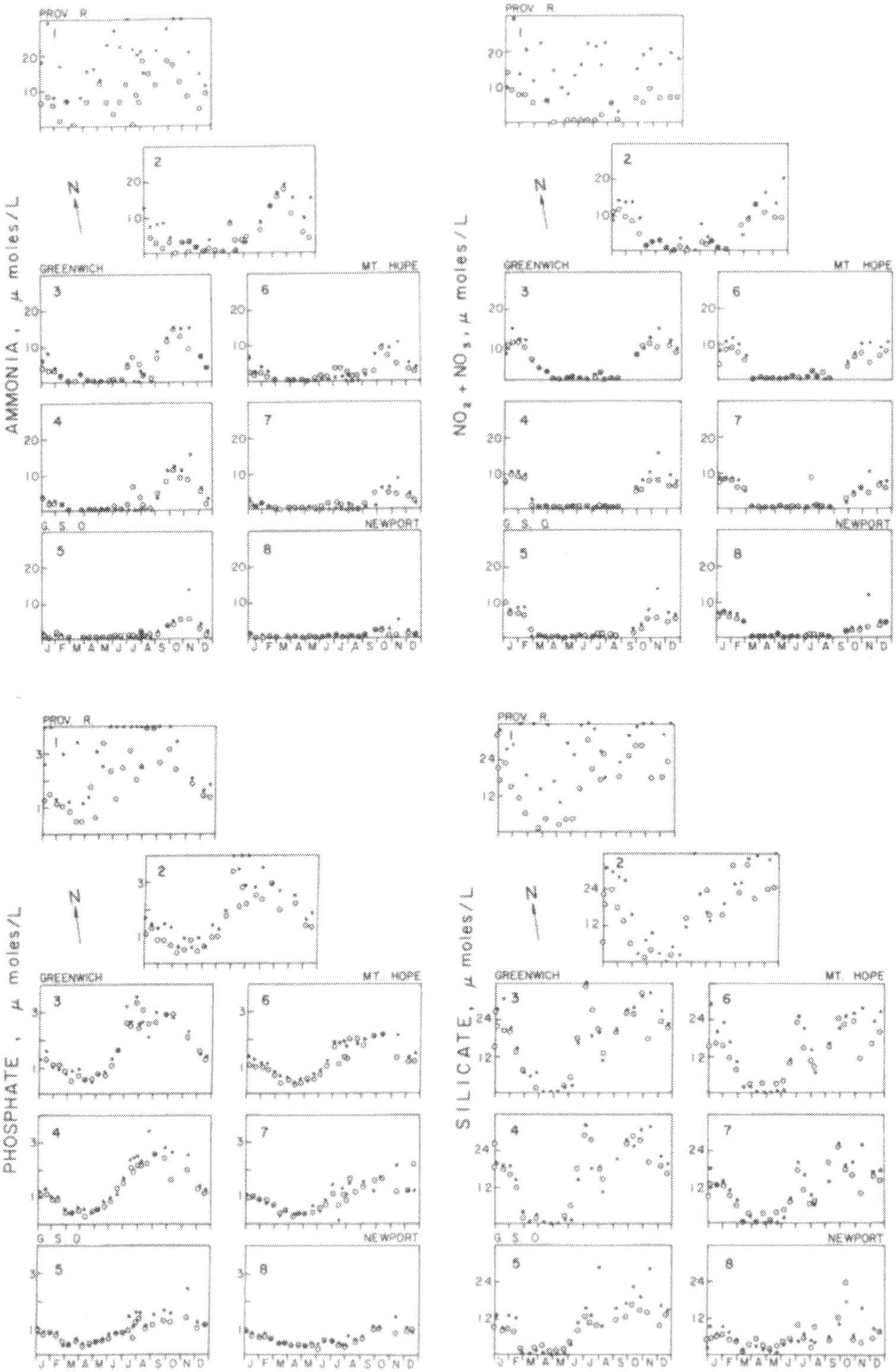

AMMONIA, μ moles/L
NO₂ + NO₃, μ moles/L
PHOSPHATE, μ moles/L
SILICATE, μ moles/L
PROV. R.
GREENWICH
MT. HOPE
G. S. O.
NEWPORT
N
J F M A M J J A S O N D

Fig. 4

is about an $8^0/_{00}$ range from salinities of about $24^0/_{00}$ in the Providence River to a maximum of 32 or $33^0/_{00}$ at the mouth of the bay. In general, there is no well-defined halocline except in the upper bay and river. Seasonal variations in salinity are slight except in the uppermost portion of the system.

There are strong seasonal cycles and sharp gradients in the distribution of biologically important nutrients, including ammonia, nitrite, nitrate, phosphate, and silicate (Fig. 4). Maximum concentrations are found in the upper bay and river and reflect the large inputs of sewage in this section. The sewage effluents are also a major source of input for petroleum hydrocarbons and heavy metals. Analysis of the effluent from the largest treatment plant by Farrington and Quinn (1973) indicated that from 0.4–2.0 metric tons of petroleum hydrocarbons per day enter the bay near the city of Providence from that one plant alone. Measurements of heavy metals in the effluent from a smaller treatment plant on the bay have been given by Ryther et al. (1972).

1.3.2 Biology

Narragansett Bay is a phytoplankton-based ecosystem in which water depths and turbidity, as well as a lack of firm substrate, have minimized the importance of attached algae and vascular plants. The phytoplankton populations have been described by Smayda (1957, 1973a), and by Pratt (1959, 1965). In general, the seasonal cycle is characterized by much greater standing crops than are found in adjacent waters and by a marked winter–early spring bloom that preceeds a series of intense summer blooms (Fig. 5). The winter population is usually dominated by the diatoms *Skeletonema costatum*, *Thalassiosira nordenskiöldii*, *Asterionella japonica* and *Detonula confervacea*, while the summer flora is composed predominantly of flagellates such as *Olisthodiscus luteus* and microflagellate species. The winter bloom appears to begin in the upper West Passage, and then spreads or is carried throughout the bay. While there is a sharp increase in the metabolism of the plankton community associated with this bloom, algal production levels are also high in summer during a second bay–wide bloom. Occasional localized blooms with large standing crops and high production are also found associated with periods of heavy runoff or other disturbances in spring and summer.

The major consumers of the bay are the zooplankton, which are dominated alternately in winter and summer by the ubiquitous copepods, *Acartia clausi* and *A. tonsa*, which may make up 95% of the total population (Frolander, 1955; Martin, 1965; Jeffries and Johnson, 1973). Zooplankton biomass is greatest in early summer (Fig. 6), and Martin (1968) has suggested that the increasing grazing pressure during the spring terminates the winter phytoplankton bloom in the bay. His preliminary

◀

Fig. 4. Surface (●) and bottom (○) concentrations of ammonia, nitrite plus nitrate, phosphate and silicate in Narragansett Bay from Aug. 1972–Aug. 1973. Measurement were taken at 13 stations around the bay and averaged for each of the eight spatial elements of the ecosystem model (see Fig. 1; data from Narragansett Bay Systems Ecology Project, Nixon et al., in preparation)

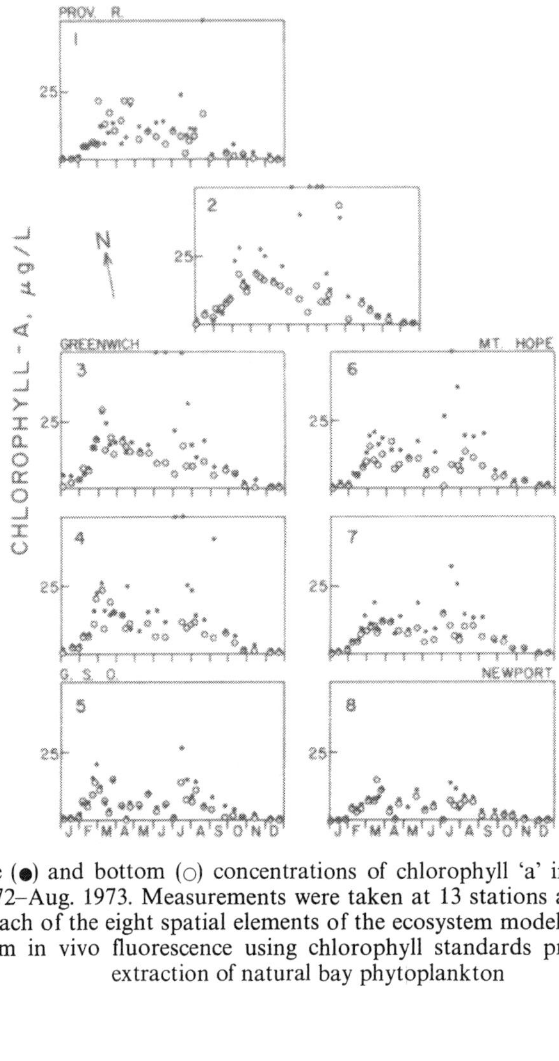

Fig. 5. Surface (●) and bottom (○) concentrations of chlorophyll 'a' in Narragansett Bay from Aug. 1972–Aug. 1973. Measurements were taken at 13 stations around the bay and averaged for each of the eight spatial elements of the ecosystem model (see Fig. 1). Values calculated from in vivo fluorescence using chlorophyll standards prepared by acetone extraction of natural bay phytoplankton

work and the recent detailed investigations by Vargo (1976), suggest that the excretion of nitrogen by the zooplankton may be important in providing this nutrient for the summer phytoplankton.

The zooplankton are subjected to predation from a number of sources. Fish larvae and meroplankton occur throughout the bay in an irregular pattern during spring and summer (Herman, 1958; Matthiessen, 1974). Dense populations of the carnivorous ctenophore, *Mnemiopsis leidyi*, begin to develop rapidly in the upper bay and Providence River early in the summer, then spread throughout the bay (P. Kremer and Nixon, 1976). Maximum biomass is usually found during August, following a sharp decline in the smaller zooplankton (P. Kremer, 1976). Commercially important stocks of Atlantic menhaden, *Brevoortia tyrannus*, migrate into Narragansatt Bay at various times during the summer. While the juveniles

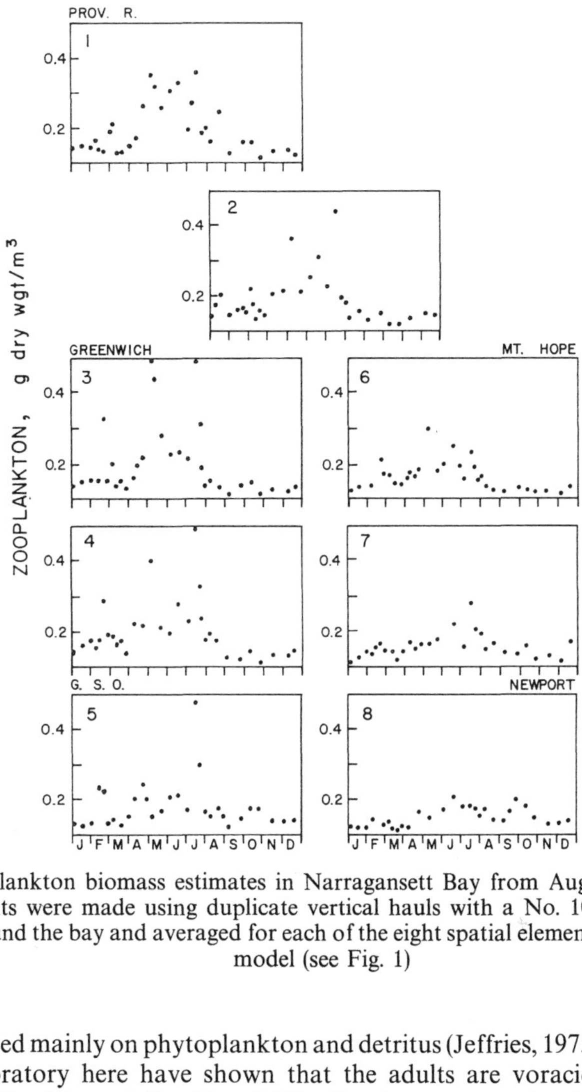

Fig. 6. Zooplankton biomass estimates in Narragansett Bay from Aug. 1972–Aug. 1973. Measurements were made using duplicate vertical hauls with a No. 10 (150 μ) net at 13 stations around the bay and averaged for each of the eight spatial elements of the ecosystem model (see Fig. 1)

appear to feed mainly on phytoplankton and detritus (Jeffries, 1975), feeding studies at the laboratory here have shown that the adults are voracious zooplankton predators (Durbin and Durbin, 1975). Carnivores may also influence the bay through nutrient excretion. While calculations indicate that this is probably a very minor impact of the menhaden (Durbin, 1976), nitrogen excretion by the ctenophores may be comparable to or exceed that of the smaller zooplankton (P. Kremer, 1975). Higher-level trophic interactions have been less intensively studied. Predation by butterfish, *Preprilus triacanthus*, appears to play a role in causing the fall decline of the ctenophores (Oviatt and Kremer, 1977) and the bay is characterized by abundant populations of carnivorous sport fish such as striped bass (*Morone saxatilis*) and blue fish (*Pomatomus saltatrix*) that feed on the menhaden. The demersal fish of the bay are abundant with a relatively constant biomass of some 0.05 g m^{-2} dry weight that is largely sustained by feeding on the infauna and epifauna of the sediments (Oviatt and Nixon, 1973).

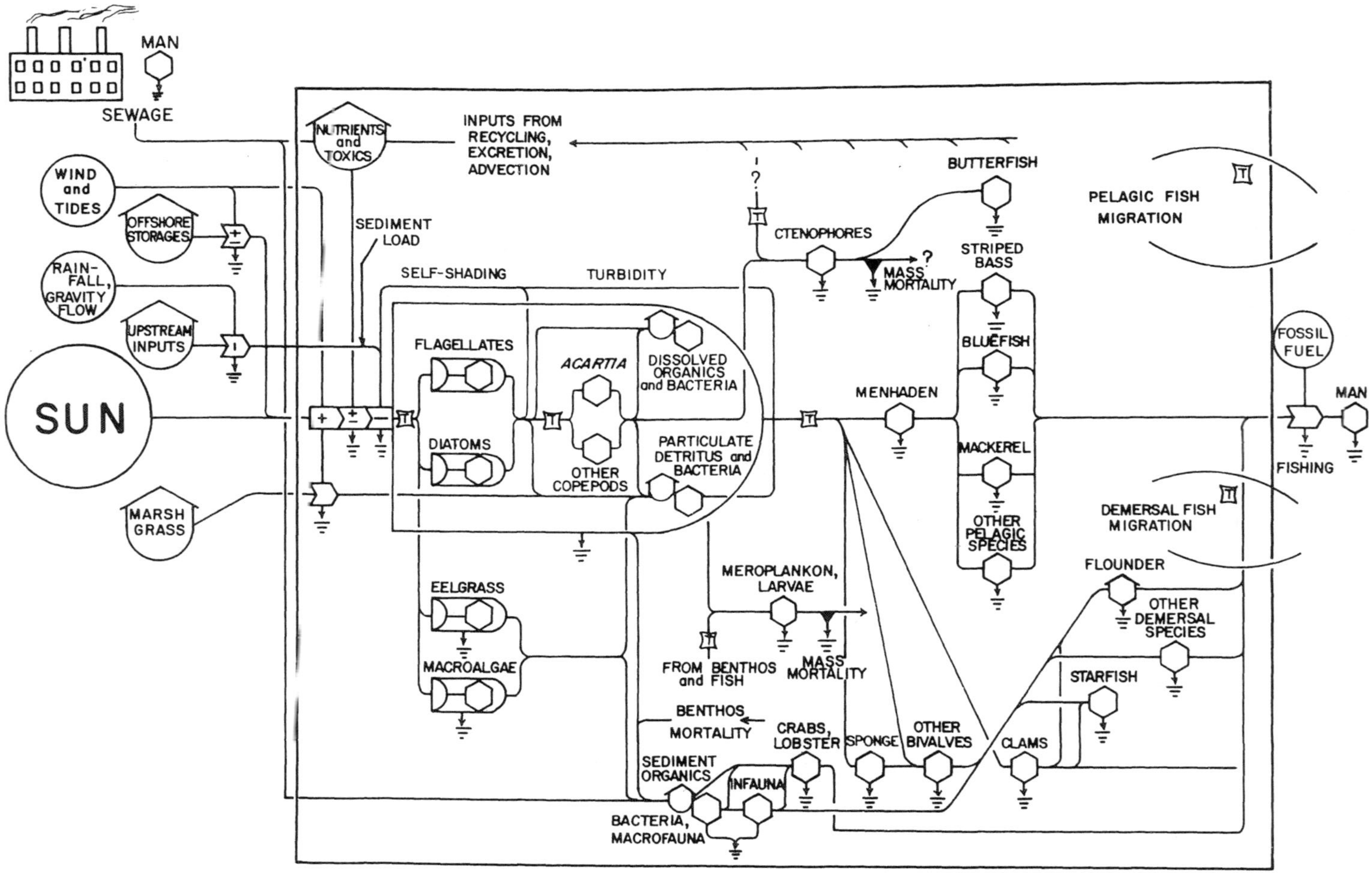

Fig. 7. A complex, but still greatly simplified energy flow diagram for the Narragansett Bay ecosystem on a summer day. This conceptual model was a first step in the process of abstraction that led to development of the numerical model. Symbols follow Odum (1972), and have been used to synthesize past and on-going bay studies by a large number of people

The community of larger benthic animals, however, is not dominated by the fin fish, but by dense populations of the hard clam, *Mercenaria mercenaria*. These animals appear to have a standing crop of some 0.6 g m^{-2} dry meat weight (Russell, 1972) and can exert a strong grazing pressure on the phytoplankton except when water temperatures are low (Loosanoff, 1939). The bay bottom consists largely of empty clam, mussel, scallop, and oyster shells, with an epifauna of starfish, lobsters, conchs, scattered sponge beds, crabs, and swarms of grass shrimp and sand shrimp. There is an abundant infauna that has been described by Stickney and Stringer (1957) and by Phelps (1958). The major communities are deminated by the bivalves *Nucula proxima* and *Yoldia limatula* and the polychaete *Nephthes incisa*, with densities of about 8.0 g m^{-2} dry weight.

On the basis of the standing crop maintained, it appears that much of the high-phytoplankton production of the bay is directed to benthic food chains and the support of an abundant infauna and large populations of bivalve molluscs and flounder. A preliminary energy flow diagram for the Narragansett Bay ecosystem indicates the interrelationships of these flows and storages (Fig. 7). Inputs of energy from marsh detritus and sewage may also be important in the upper bay. Studies of marsh production along the bay (Nixon and Oviatt, 1973, 1973a) indicate a potential total input of detritus of $1.9 \cdot 10^6$ kg $year^{-1}$. The secondary production of pelagic bacterial biomass in the bay also appears substantial in preliminary calculations by Sieburth (personal communication) and may serve as an important food source for some of the smaller species.

2. The Narragansett Bay Model

2.1 General Formulation Strategy

One of the basic problems in formulating mathematical expressions for detailed ecological processes is how to combine into a single expression the influences of several factors that operate simultaneously. The approach chosen here, and in many other similar models (Canale et al., 1974; DiToro et al., 1971; O'Brien and Wroblewski, 1972), is to postulate a maximum rate as a function of one factor, usually temperature, and to determine the effects of the remaining factors as unitless fractions which reduce the maximum. Functionally, this is the same assumption implicit in commonly used equations for even a single factor. For example, the hyperbolic Monod (1942) expression for substrate-limited growth of microorganisms (also used in enzyme reaction kinetics by Michaelis and Menten, 1913) is the product of a maximum growth rate, μ_{max}, and a fraction:

$$\mu = \mu_{max}\left(\frac{S}{K_S + S}\right).$$

Note that numerator and denominator have the same units (concentration), and the fraction is therefore a unitless number between 0 and 1. This logic may be easily extended to include more limiting fractions, each of which may be independently derived.

This approach is used a number of times in our model. Phytoplankton growth is represented as a temperature-dependent maximum, reduced by less-than-optimal light and limiting nutrients.

$$\mathrm{PG}(\mathrm{mgC\,l^{-1}\,day^{-1}}) = \mathrm{PG_{max}}(\mathrm{mgC\,l^{-1}\,day^{-1}}) \cdot \mathrm{LTLIM} \cdot \mathrm{NUTLIM}.$$

Similarly, grazing by copepods and fish larvae are determined by a temperature-dependent maximum preferred ration, and reduced by density-dependent food limitation. Food preference indices might easily be added as additional fractional terms.

2.2 A Conceptual Overview

The goal of the ecological model is to depict the major features of both the spatial variations throughout the bay and the temporal changes during the year for the dominant plankton system of Narragansett Bay. The previous chapter presented a brief description of the complex natural ecosystem of the bay. The first step in the modeling process is to develop a conceptual model by generalizing those aspects considered to be of primary interest, and simplifying or even omitting properties believed to be of secondary importance. This is a critical step, for it is the features composing this analog that are ultimately the sole basis for interpretation of the model. These features, for our purposes, include phytoplankton, smaller, primarily herbivorous zooplankton, and the chemical nutrients ammonia, nitrate and nitrite, phosphate, and silicate.

These are the six state variables of the model that are represented in substantial detail to allow critical evaluation of the overall patterns of abundance, and the numerous rates and exchanges within and between the components. Of course, it is not meaningful to attempt such a detailed treatment without immersing these state variables in a system that represents the effects of other factors. Thus the conceptual model was expanded to include those secondary compartments that were anticipated to be important influences on the plankton-nutrient system. Within the model, these secondary compartments are represented by forcing appropriate patterns with little mechanistic detail, except as thought necessary to express important feedbacks and internal controls. In most cases these compartments of the model are formulated simply, as much for lack of necessary detailed information as for any other reason.

The simplified conceptual model in Figure 8 schematically represents the mass and energy flows and controlling interactions among the compartments of the Narragansett Bay ecosystem model. At this point, a brief synopsis of the processes that have been considered will serve as an introduction to the detailed formulations presented in the following chapters.

The processes of physical circulation in major estuaries such as Narragansett Bay make it difficult to investigate the dynamics of the ecological system without a hydrodynamic model that simulates at least the major tidal features. The representation of tidal mixing in the ecological model is based on a numerical hydrodynamic model of Narragansett Bay developed at the University of Rhode Island as part of the same Sea Grant program that supported this research. A crude but efficient scheme driven by daily variations in tide height at Newport was developed, which represents the net transfer of chemical and biological concentrations throughout the geographical regions of the bay. For this purpose, the bay is divided into eight spatial regions, or elements, which are assumed to be both vertically and horizontally homogeneous. Each of the elements retains characteristics of depth, volume and seasonal temperature appropriate to the location in the bay. The biological and chemical components of the system function within each element on a daily basis, with tidal processes resulting in circulation among appropriate regions. River flow is programmed to follow a seasonal cycle in agreement with long-term observations.

Solar radiation is the driving energy source for the phytoplankton, and thus for the rest of the biological system. For this reason considerable attention was paid to developing an accurate yet flexible representation. Because a daily time base was used for all rates, total insolation per day was chosen as the unit of solar input. The diel irradiance pattern was not ignored however, and its effect on phytoplankton

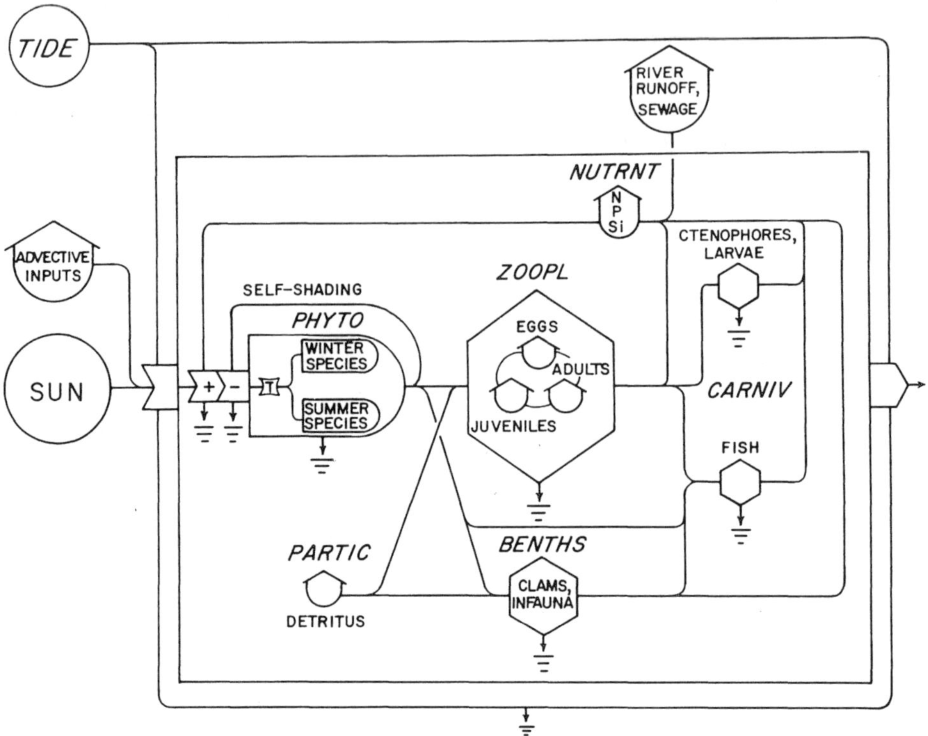

Fig. 8. Final energy flow diagram and the conceptual framework for the numerical model of the Narragansett Bay ecosystem. While much of the complexity of Figure 7 has been omitted, some detail has been added in the zooplankton compartment. Only the phytoplankton, zooplankton, and nutrient compartments are fully simulated with mechanistic detail. This same conceptual model is used in each of the eight elements of the bay (Fig. 1), with appropriate values for each compartment and forcing functions. Names of computer program subroutines in italics

production was accounted for. Seasonal variations in insolation may be introduced in the program by any of three options: (1) as smooth curve linearly interpolating between monthly means, (2) actual data values for a given year, or (3) a stochastic scheme representing reduction of a theoretical maximum due to daily variable cloudiness.

Phytoplankton are the primary producers of the system. The basis of the formulation is a maximum potential growth, calculated as a function of temperature, with reduction in this rate due to less than optimum light and insufficient nutrients. The light factor is determined from a time and depth integration of a photosynthesis–light response including surface inhibition. The optimal insolation for photosynthesis acclimates to the changing light regime with a short time-lag, though a lower limit to this acclimation may be imposed. Self-shading by the phytoplankton is included as a component of the extinction coefficient. Nutrient limitation is determined from a hyperbolic Monod calculation assuming a "most limiting" nutrient.

Ammonia and oxidized forms of nitrogen are considered as well as phosphate and silicate, and the nutrient selected as limiting may vary both temporally and spatially. Preference for ammonia is represented simply, by its being taken up before nitrate plus nitrite concentrations are utilized. A number of options are provided in order to make the phytoplankton compartment more useful in hypothetical analyses. Two species groups are provided for, which may be assigned different temperature or nutrient responses. Both the kinetic half-saturation constants and the nutrient content of the two groups may be varied. While the specification of two groups was originally designed for diatom vs. dinoflagellate dynamics, the formulations are general enough to represent a variety of the ways in which phytoplankton communities differ. Finally, the model accepts a range of nutrient-to-carbon ratios for each species group and nutrient. A simplified scheme, allowing the ratios to vary within the specified ranges has been developed to represent the competitive advantage of luxury uptake.

The zooplankton compartment depicts the community of primarily herbivorous, estuarine copepods found in the bay. Formulations were based largely on available information about the *Acartia* species, which usually dominate. The compartment is subdivided into adults, eggs and juveniles. Adults consume available food—phytoplankton, detritus, and their own eggs and juveniles—in a density-dependent manner and as a function of temperature. If the assimilated ration exceeds the temperature-dependent respiration, eggs are released. Nutrient excretion occurs in proportion to respiration, after a correction for changes in the metabolic substrate being utilized by the adults and juveniles.

Eggs hatch after a delay determined by ambient temperature, and juvenile development ensues. Development is also controlled by temperature, with the effect being mediated through commensurate changes in maximum growth rate and development time. Thus the emergence of the equivalent of one adult copepod after the development of the equivalent of one hatching egg is assured, despite the fact that the model maintains only homogeneous carbon pools, rather than numbers of individuals. Food limitation, respiration and assimilation all enter into the calculation of the realized daily growth rate. Predation by carnivores depletes adults, eggs, and juveniles, while the additional pressure of adult cannibalism further reduces eggs and juveniles. Juveniles may be excluded from adult predation during a portion of their development, however, assuming that cannibalism on near-adults may be reduced. The unassimilated fraction of ingestion is transferred to the benthos, where its contribution toward supporting nutrient fluxes is assessed.

The benthos in the model affects the plankton system in two ways. Based on the abundance of hard clams in the bay, the impact of bivalve filtering on the standing crop of phytoplankton is included. Because of considerable uncertainty in their role, however, the formulation is quite general. A lower temperature limit to feeding may be set, above which filtration increases with a specified Q_{10} up to a maximum rate. The clam biomass in each element reflects the estimated populations around the bay.

The second role of the benthos in the model concerns the regeneration of nutrients. Fluxes of ammonia, phosphate, and silicate are forced into the water column as a function of temperature. Technically, the nutrient cycles in the model are uncoupled at this point—the fluxes occur according to empirical regressions without regard to the ultimate balance of sources. However, several contributors of nutrients are accounted for, including copepod unassimilated fecal material, sinking diatoms, and clam ingestion. The net balance of these sources against the forced fluxes is determined, despite the lack of direct mechanistic feedback.

Particulate organic matter undoubtedly plays an important role in the bay system. Unfortunately, the sparcity of data on the concentration and composition of particulates precluded adequate evaluation of this in the model at this time. The potential role in some areas was anticipated, however, and a subroutine has been included that provides a vehicle for implementation of future observations. Presently, particulate carbon concentration must be forced—i.e., with no specification of sources, etc.—and it is assumed to be available food for the copepod population. Many obvious improvements in this compartment are contemplated including, for example, the coupling of fecal pellet generation to the particulate compartment, and possible feeding preference indices, which would allow particulate carbon to be utilized as a nutritional resource only when phytoplankton were relatively unavailable. Preliminary indications of the model suggest that some additional food is probably available to the zooplankton and the benthos, at least for parts of the year.

The seasonally important predation pressure of three carnivores on the zooplankton is also represented in the model. The ctenophore, *Mnemiopsis leidyi*, becomes an important predator during the late summer months. Feeding and nitrogen excretion are calculated in proportion to the biomass interpolated daily from monthly population estimates throughout the bay. Also during the summer and fall, the Atlantic menhaden, *Brevoortia tyrannus*, invades the bay in great numbers. Based on initial estimates of population size and metabolic requirements, a preliminary formulation of their potential grazing impact was developed. Consideration of a lower-size threshold for menhaden plankton filtering, combined with the observation that small flagellates usually dominate the region of the bay when and where the fish are found, suggested that their role as carnivores on the zooplankton was more significant than the more traditional herbivorous mode. Thus the model includes menhaden feeding on zooplankton in the upper reaches of the bay, using a crude estimate of a maximum daily ration and food-limitation term, in a form similar to that of the zooplankton.

The last component of carnivores is fish larvae. An extensive survey of Narragansett Bay provided detailed seasonal population estimates. In a formulation similar to that for menhaden, feeding is determined as a food-density-dependent function of forced biomass levels around the bay. In this case, the maximum ration of the fish larvae is temperature-dependent, since their presence in the bay spans a longer time period, and thus a greater temperature range, than when either the ctenophores or the menhaden are abundant. Grazing for all the carnivores is combined as a single estimate of an instantaneous filtering rate, affecting the eggs, growing juveniles, and adults of the zooplankton compartment.

The final model compartment, and the one which provides a critical feedback loop interconnecting all components of the bay ecosystem, is that of chemical nutrients. Excretion by animals, uptake by algae, and fluxes from the benthos are contributions to the nutrient balances which are determined by the dynamics of the respective compartments. In addition, the input of nutrients in domestic and industrial sewage entering throughout the bay is included, based on flow estimates and concentration analyses on samples collected seasonally from the major treatment plants on Narragansett Bay. Phosphate-phosphorus and silicate-silica are the forms of these nutrients throughout the model. Ammonia and nitrate-plus-nitrite nitrogen concentrations are maintained, with nitrification of the reduced nitrogen into oxidized forms proceeding at a temperature-dependent daily rate. No organic nutrients are presently included in the analysis, and future improvements may require the addition of these forms to represent the nutrient budget for the bay more completely.

Theoretical Formulations

3. Physical Forcing Functions

3.1 Temperature

In a temperate estuary like Narragansett Bay, one of the most important forces operating on the ecological system is temperature. The major variation is seasonal, although some diel and spatial fluctuations also occur. Within the model, diel and vertical temperature differences are ignored while temperature varies among the spatial elements on a daily schedule. The observed seasonal pattern closely approximates a sinusoidal curve, and a simple equation may be formulated which matches the time of maximum or minimum temperature and the yearly amplitude. The seasonal pattern of the baywide average of all temperature measurements during the field sampling program can be described with the equation:

$$\text{Temp}(^\circ\text{C}) = 11.5 - 8.5 \cos\left[2\pi(\text{day} - 40)/365\right]. \tag{1}$$

As the Julian day goes from 1 to 365, the cosine goes from -1 to $+1$, and the temperature oscillates around the mean 11.5° C with an amplitude of 8.5° C (Fig. 9).

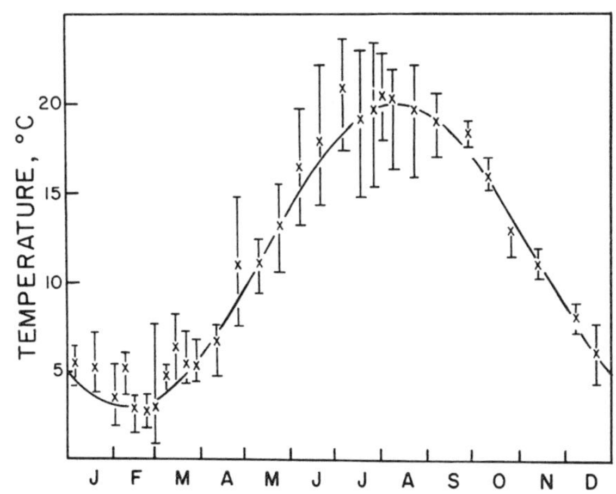

Fig. 9. Mean and range of surface and bottom water temperatures at the 13 stations sampled from Aug. 1972–Aug. 1973 (Fig. 1). *Smooth curve:* average bay temperature calculated from Eq. (1). A correction factor for each of the eight elements is applied to the calculated mean value to account for spatial variation within the bay (Table 1)

Table 1. Temperature deviations from the baywide mean predicted by the equation $T = 11.5–8.5 \cos[2\pi(\text{day-}40)/365]$ used in the ecological model. Based on sampling year Aug. 1972–Aug. 1973

Month	Element							
	1	2	3	4	5	6	7	8
JAN	−0.6	−0.6	−1.0	−0.3	+0.6	−0.2	+0.4	+1.0
FEB	−0.1	0.0	−0.3	−0.3	−0.5	0.0	+0.2	−0.6
MAR	+0.6	+0.3	+0.5	+0.3	−0.5	0.0	−0.5	−0.2
APR	+0.7	+0.5	+0.8	+0.4	−0.8	0.0	−0.7	−1.6
MAY	+0.8	+0.8	+0.8	+0.4	−0.8	+0.4	−0.6	−1.6
JUN	+1.5	+2.0	+0.8	+0.4	−1.2	+0.3	−0.8	−2.9
JUL	+1.2	+1.6	+1.0	+0.4	−0.5	−0.3	−1.0	−2.5
AUG	+0.8	+1.0	+1.0	+0.5	−0.5	+0.5	0.0	−2.4
SEP	+0.4	+0.2	+0.2	+0.5	−0.3	+0.4	+0.1	−1.2
OCT	0.0	0.0	0.0	+0.2	−0.2	0.0	+0.1	0.0
NOV	−0.2	−0.4	0.0	0.0	−0.1	0.0	0.0	+0.4
DEC	−0.3	−0.6	−0.5	0.0	+0.2	−0.1	−0.2	+0.7

To allow for spatial variations, monthly deviations from the baywide mean were estimated for each of the eight elements throughout the year (Table 1). Thus, for each day the temperature in each spatial element is determined by adding the appropriate monthly deviation to the baywide mean of Eq. (1).

3.2 Solar Radiation

To evaluate the role of light acclimation by the phytoplankton, it was desirable to formulate a scheme for incident solar radiation that incorporated day-to-day variations, rather than simply a smooth average curve for the seasonal trend. While daily radiation data for the Narragansett Bay region are available through the Eppley Laboratory in Newport, R. I., and may be used directly in the model, a more general approach was also developed. For the general case, a theoretical maximum radiation equation was determined for this latitude, which was combined with average cloud-cover data to simulate daily received radiation, or insolation, as described below.

Based on navigation tables (Bowditch, 1958), the sun angle was determined at hourly intervals for the summer and winter solstices and the spring equinox. A meterological equation was then used to predict the radiation received on a flat surface as a function of sun angle (h) for a clear sky (Sverdrup et al., 1942).

$$I_{\text{clear}} = S\,\mathrm{e}^{-Ta_m m}\sin(h) + D, \tag{2}$$

where: S = solar constant (2.0 ly/min),

T = atmospheric turbidity factor,

m = relative air-mass length $= \sin(h)^{-1}$ atmospheres,

a_m = clear sky extinction coefficient = 0.128–0.054 log(m), atm^{-1},

h = sun angle, degrees,

sin(h) = corrects the calculation from a normal to a horizontal surface,

D = diffuse radiation component, as a function of sun angle: $D = 0.44 \exp(-0.22h)$ [fitted by least square regression to data in Reifsnyder and Lull (1965)].

The hourly rates from this equation were then integrated to provide estimates of the total insolation during the daylight hours. In addition, the daylength or photoperiod, f, was determined directly from the sun angle calculations. A simple sinusoidal function was determined, similar to that for temperature, which fit the predicted clear-sky maximum at the solstices, and a second equation was developed for the photoperiod:

$$\mathrm{RADN}_{\max} = 677.5 - 371.5 \cos[2\pi(\mathrm{day} + 10)/365] \quad (3)$$

$$f = 0.5 - 0.125 \cos[2\pi(\mathrm{day} + 10)/365]\,. \quad (4)$$

The success of the equation in predicting the calculated equinox reflects the accuracy of the sinusoidal assumption:

Date	Insolation. (ly/day)		Photoperiod Eq. (4)
	Calc.	Eq. (3)	
June 21–Summer Solstice	1049	1049	0.625
March 21–Spring Equinox	696	669	0.500
December 21–Winter Solstice	306	306	0.375

The predicted clear-sky maximum may then be adjusted for cloudiness using the equation (Sverdrup et al., 1942):

$$\mathrm{RADN} = \mathrm{RADN}_{\max}(1.0 - 0.071\,C)\,, \quad (5)$$

where: C = cloud cover (tenths).

The New England Weather Summaries (U.S. Weather Bureau) report daily estimates of cloud cover for the weather station at Green Airport, Warwick, R. I. Fifteen years of daily data (1959–1973) were tabulated, and normal monthly patterns were characterized. Interestingly, the distributions are strikingly bimodal for clear or cloudy days (Table 2) with almost all months showing an abundance of cloudy days ($C = 10$). In the model, an IBM-supplied subroutine (RANDU, Scientific Subroutine Package) generates random numbers that are adjusted to conform to these monthly distribution patterns.

The result of this scheme overestimates the radiation, however, and must be corrected for extinction by other atmospheric materials in addition to cloudiness.

Table 2. Monthly distribution of cloud cover for the Narragansett Bay region[a]. Percent of days with daytime cloud cover in tenths

	Cloud cover, tenths										
	0	1	2	3	4	5	6	7	8	9	10
JAN	15.5	7.7	7.3	4.5	4.7	6.2	5.4	5.6	8.0	7.1	28.0
FEB	13.2	5.7	4.5	4.7	6.1	5.9	7.8	6.6	7.5	9.9	28.1
MAR	9.5	5.2	6.7	6.0	6.4	5.6	5.8	6.9	7.1	9.4	31.4
APR	7.3	5.8	6.9	7.1	7.6	6.2	7.1	7.3	8.7	12.2	23.8
MAY	6.7	4.6	5.1	7.1	5.5	9.0	8.7	8.8	12.0	9.0	23.5
JUN	4.1	5.2	6.0	7.1	9.0	9.3	10.5	6.2	9.3	12.4	20.9
JUL	1.9	6.2	8.0	7.1	9.0	8.2	7.1	11.0	8.4	14.8	18.3
AUG	2.2	8.4	8.4	9.9	7.1	9.0	9.0	9.5	11.8	9.2	15.5
SEP	8.7	8.0	6.6	10.2	7.1	6.4	7.5	6.9	8.2	8.9	21.5
OCT	13.1	8.2	9.2	6.0	6.9	5.4	6.7	6.9	8.4	9.4	19.8
NOV	7.1	6.9	7.1	5.8	5.5	2.9	5.1	5.8	8.9	12.2	32.7
DEC	10.5	8.8	4.3	3.7	6.7	5.2	6.7	6.4	6.2	8.4	33.1

[a] Data from Local Climatological Data, U.S. Weather Service, for Green Airport weather station, Warwick, R. I. 1959–1973.

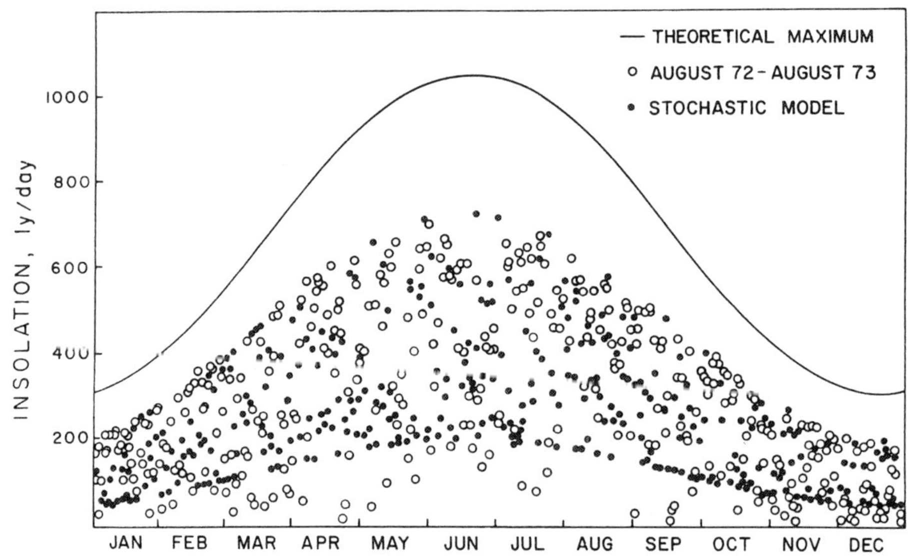

Fig. 10. Comparison of theoretical and measured insolation. Maximum clear-sky insolation for Narragansett Bay calculated from Eq. (3). Addition of a stochastic cloud-cover factor [Eq. (5), Table 2] and a correction for atmospheric attenuation brought the calculated values close to those observed by the Eppley Laboratory in Newport, R.I., from Aug. 1972–Aug. 1973

Ideally this is represented by the turbidity factor in Eq. (2). However, since no estimates of atmosphric extinction coefficient were available for the bay region, Eq. (3) was multiplied by an empirical factor of 0.7, which resulted in a lowering of the upper predictions by 30% to agree with the observed data for the year of the sampling program (Fig. 10). Lowry (1967) states that urban atmospheres reduce light by 15%, and Reifsnyder and Lull (1965) depict the effect of an industrial absorption as varying with sun-angle between 0.7 and 0.8 of the standard clear-sky irradiance. Considering industrial inputs combined with the humidity of a maritime climate, the reduction by 30% seems reasonable.

Even with this empirical turbidity correction, a number of the days actually received substantially less light than the lowest predictions of the stochastic model. An examination of the cloud-cover correction Eq. (5) reveals that even at complete cover ($C=10$), 29% of the radiation is predicted to penetrate. In practice, a cloud cover of 10.0 may be recorded for a range of cloud densities, and the inaccuracy of the relation may be greatest for heavy overcast conditions. In any case, the sensitivity of the model to unusually low or high radiation levels is easily evaluated by variations in the monthly patterns that are used in the model.

3.3 Tidal Circulation

In many estuaries, the interaction of the local tides, river flows, winds, and geography make it difficult to predict the dynamics of ecological systems without a hydrodynamic model that simulates currents, flow rates, and flushing times throughout the system. In fact, even the extent to which these processes have ecological implications is poorly known. A numerical tidal model has been developed for Narragansett Bay by Hess and White (1974) through an extension and adaptation of the basic two-dimensional long-wave propagation models of Leendertse (1967). At certain times of the year, especially in the upper reaches of the bay, vertical stratification occurs, which may have significant implications. Since most of the bay is not usually strongly stratified, however, the model is averaged in the vertical dimension using equations whose mathematical development has been summarized by Pritchard (1971). In the Hess and White model, a grid system was selected to fit the complicated geometry of the bay and to give as detailed a picture of the hydrodynamics as was practical. The result was a field with 324 operational elements each 0.5 nautical miles on a side. Basic Navier-Stokes momentum equations plus a conservation of mass equation were applied to each of the vertical cells defined by the grid system. The Chezy relationship for bottom friction and a quadratic approximation for wind stress were included. Since the model is vertically averaged, only the gross effect of wind drag on total transport is represented. The driving forces at the boundaries consist of the astronomical tide at the mouth, the discharges from the two largest rivers, and the tidal velocity at the mouth of Mt. Hope Bay (Fig. 1).

The hydrodynamic model has been verified with field observations of tide heights at Newport, Bristol and Providence where data are available. Current velocity observations from near-surface and near-bottom meters and from drifting poles were also available for comparison with model outputs in the lower East and

West Passages (Levine, 1972; Weisberg and Sturges, 1973). The computed velocity compares favorably with that observed (Hess and White, 1974). In addition, the model has been applied to the historical simulation of hurricances in the bay, where damage was so great in the storms of 1938, 1944, and 1954 that the Army Corps of Engineers proposed construction of an extensive series of barriers across the bay (Hicks et al., 1956). Comparison of the results of the simulated and real storms in affecting surge at Providence was also quite convincing.

The extensive detail of the grid scheme and the computational timestep of about four minutes result in an execution time for the computer of ten minutes or longer for a simulation of one tidal cycle. For the ecological model, where long real-time simulations are needed and the rate constants are determined for time intervals on the order of a day instead of minutes, the detailed hydrodynamic model could not be employed directly. As a working solution to this interfacing problem, the detailed bay grid was grouped into eight large elements, or ecological subsystems (Fig. 1). The advective transport of particles and dissolved materials among these large elements was estimated using the detailed model.

The chöice of the eight elements was made somewhat arbitrarily in an attempt to localize regions that might be expected to exhibit similar characteristics based on basin geometry. An additional criterion was that, when possible, boundaries were chosen where circulation patterns might be expected to be relatively uncomplicated, such as between parallel shorelines. The physical characteristics of the eight elements are shown in Table 3.

A salinity advection version of the fine-grid model was employed to develop the more simplified mixing program for the eight element divisions. All of the fine grids representing one of the large elements were assigned an initial tracer concentration of 1.0 arbitrary unit. All other grids were initialized at a concentration of zero with

Table 3. Physical dimensions of spatial elements of the Narragansett Bay ecological model (see Fig. 1)

Element	Location	Area[a] (10^7 m^2)		Volume[b] (10^8 m^3)	Depth[c] (m)
		A	B		
1	Providence R.	2.007	3.005	1.30	4.3
2	Upper Bay	4.459	4.121	3.00	7.3
3	Greenwich Bay	2.851	2.490	1.15	4.6
4	Mid-West Passage	6.190	6.526	4.63	7.1
5	Lower West Passage	2.001	2.061	2.04	9.9
6	Mt. Hope	2.599	2.833	2.22	7.8
7	Mid-East Passage	3.854	3.778	5.73	15.2
8	Newport	2.415	2.318	5.54	23.9
	Total	26.376	27.132	25.61	
	Average				10.0

[a] Area estimates: A: By planimeter from U.S.C.&G.S. Chart 353. B: From grid area of hydrodynamic model (Hess an White, 1974) 0.25 sq. naut. mi./grid.

[b] Volume from hydrodynamic model (Hess and White, 1974) for mean tide condition.

[c] Depth from model volume/area.

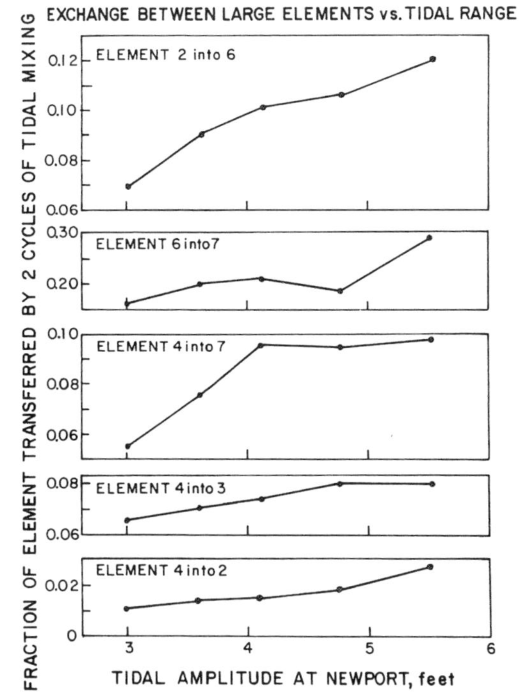

Fig. 11. Representative exchanges of water between large spatial elements of Narragansett Bay (Fig. 1) calculated as a function of tide height at Newport. Linear regressions for these and all other transfers are given in Table 4

no further input to the system. At the end of a two-tidal cycle simulation, the net daily transport to each element was calculated as a percentage of the test element volume. These calculations were made for each of the elements for five different tidal amplitudes, and regression analyses indicated that the relationships between exchange coefficients and tide at Newport were frequently linear (Fig. 11). The final eight-element tidal model is based on a set of linear equations that calculates net daily transport between any two elements as a function of tabular tidal amplitude for Newport. To insure conservation of mass, at least one exchange for each element must be calculated by difference. An algorithm was designed in which the most strongly linear regressions for each element force the computation of all remaining exchanges. All transfers affecting greater than 0.5% of the element volume were considered significant and were included. The average coefficient of determination for the 17 forced regressions used in the rapid mixing model was 0.81 (Tables 4 and 5). This scheme efficiently predicts tidal dispersion for any conservative substance throughout the eight large elements of the bay model and can handle numerous substances simultaneously. A one-year simulation requires less than 10 s of computer time.

As expected, verification of the eight-element mixing model by short runs of the parent fine-grid model was satisfactory. The major weakness is a tendency to rapidly shunt material throughout the bay due to the large size of the elements. The

Table 4. Linear regressions for net tidal exchange among spatial elements in the TIDE submodel resulting from two tidal cycles mixing, and coefficients of determination, r^2

Transfer from-into	Slope	Const	r^2
1–2	0.024	0.073	0.92
2–2	−0.038	0.897	0.94
2–6	0.919	0.018	0.94
3–2	0.026	0.083	0.88
3–3	−0.042	0.707	0.85
4–2	0.006	−0.100	0.92
4–3	0.006	0.047	0.92
4–4	−0.037	0.881	0.90
4–7	0.016	0.015	0.76
4–8	0.005	−0.015	0.79
5–5	−0.069	0.530	0.92
6–6	−0.058	0.689	0.71
6–7	0.043	0.031	0.69
7–4	0.001	0.075	0.02
7–7	0.055	0.866	0.90
8–4	0.004	−0.009	0.81
8–8	−0.097	0.716	0.96
Average			0.81

The results are based on five simulations of a detailed hydrodynamic model by Hess and White (1974) for tides at Newport from 3 to 5.5 ft. Predictions represent the fraction of one element found in another one day later.
Example: Tide height range at Newport = 5.0 ft, $T_{1-2} = 5.0 \times 0.024 + 0.073 = 0.193$ i.e., 19% of element 1 was found in element 2 after two tidal cycles of mixing.

concentrations resulting from this artifact are extremely low, however, and if results are rounded to appropriate levels of precision the results are quite reasonable.

The fast-mixing scheme was compared to earlier dye studies carried out by the U.S. Army Corps of Engineers on a hydraulic model of Narragansett Bay (U.S. Army, 1959). Two of the earlier experiments were simulated—a rapid release of dye into the Providence River (Element 1) and a ten-day continuous release into the middle-west passage (Element 4). In both cases, the agreement is satisfactory considering the widely varying methods (Fig. 12). It is especially significant that the times of peak concentration, and the slopes of the decay which represent the flushing rate of the elements agree well. While it is perhaps uncertain which of the models more accurately represents the real world, their convergence is encouraging.

3.4 Exchange with Rhode Island Sound

A serious uncertainty exists concerning the transfer of substances at the boundaries of both the real and the model Narragansett Bay. For example, it is not at all clear how much of the water that passes out of the bay during the ebb tide returns on the next flood. For the up-bay boundaries at the head of the Providence

Table 5. Average net exchange among spatial elements in TIDE submodel resulting from two tidal cycles mixing

Out of element		Into element 1	2	3	4	5	6	7	8
1	$\bar{x}$	0.825	0.175	—	—	—	—	—	—
	σ	0.040	0.025						
2	$\bar{x}$	0.062	0.738	0.062	0.023	—	0.097	—	—
	σ	0.005	0.038	0.005	0.006		0.019		
3	$\bar{x}$	—	0.193	0.529	0.277	—	—	—	—
	σ		0.027	0.045	0.021				
4	$\bar{x}$	—	0.016	0.074	0.725	0.076	—	0.084	0.004
	σ		0.006	0.007	0.039	0.014		0.018	0.005
5	$\bar{x}$	—	—	—	0.174	0.239	—	—	—
	σ				0.041	0.070			
6	$\bar{x}$	—	0.104	—	—	—	0.446	0.209	—
	σ		0.015				0.067	0.050	
7	$\bar{x}$	—	—	—	0.080	—	0.076	0.634	0.163
	σ				0.007		0.014	0.057	0.025
8	$\bar{x}$	—	—	—	0.007	—	—	0.179	0.308
	σ				0.004			0.018	0.097

Mean of five tide ranges from 3–5.5 ft at Newport, and standard deviation. Transfers are fractions of element found in other elements one day later. Based on simulations of a detailed hydrodynamic model (Hess and White, 1974).
Example: 82.5 ($\pm$4.0)% of element 1 remains after two cycles; 17.5($\pm$2.5)% of element 1 moves to element 2. All exchanges of about 0.5% or more of an element are represented.

River and the mouth of Mt. Hope Bay, the simple assumption that the returning water has the same concentration as that leaving is perhaps sufficient. At the mouth of the East and West Passages this is surely not the case. No theoretical or experimental evidence exists that contributes much to the question. Hess and White assumed 50% return in a dye-simulation with the fine-grid model (personal communication). Because preliminary indications suggested this was a factor of some significance, attempts were made to estimate at least the probable range of values. A number of approaches were tried.

1. Although the flushing time of the bay is quite uncertain—one set of calculations ranges between 10 and 60 days (Fish, 1953)—a 30-day estimate would suggest that about 1/30 of the volume leaves the bay daily or $0.085 \cdot 10^9$ m^3/day. For an average river flow of $0.5 \cdot 10^7$ m^3/day, the difference of $0.08 \cdot 10^9$ m^3/day must be lost by tidal action. This loss must be partitioned between the West and East Passages. If the loss splits in a proportion similar to the nontidal currents, 40% leaves from the West and 60% from the East Passage (Hess and White, 1974, Fig. 18B). Of the $0.08 \cdot 10^9$ m^3 loss, 60%, or $0.048 \cdot 10^9$ m^3, leaves through the East Passage, and the remaining $0.032 \cdot 10^9$ through the West Passage. Finally, the exchange regressions derived for the large element model suggest that for an

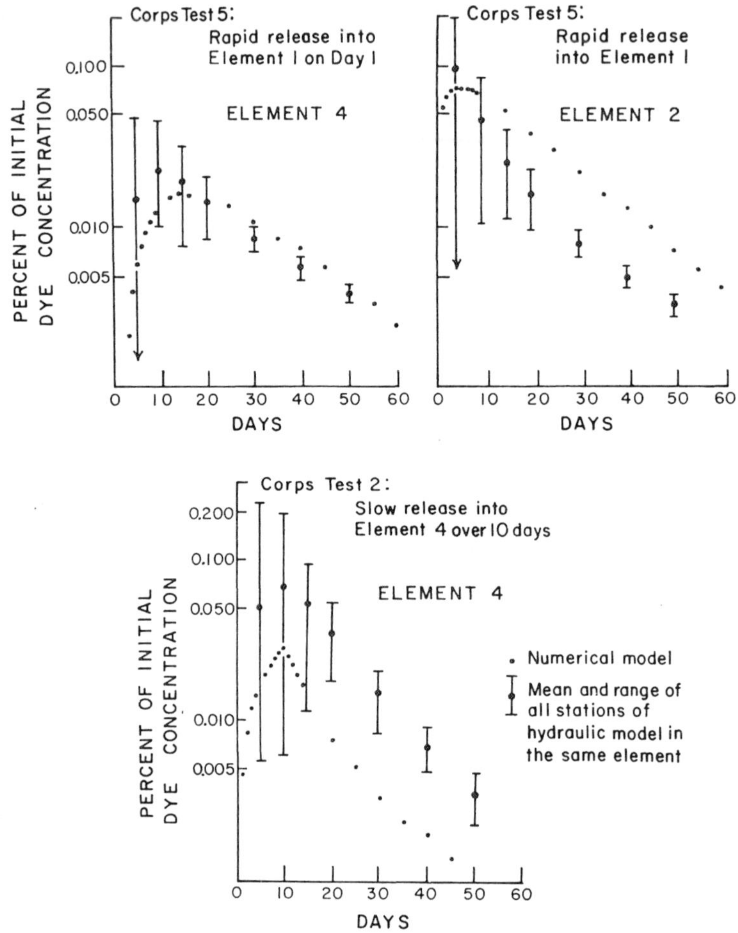

Fig. 12. Comparison of simulated dye concentrations predicted by the large-element numerical mixing model and observed concentrations in the same element area in a U.S. Army Corps of Engineers (1959) hydraulic model of the bay

average tide, $0.115 \cdot 10^9$ m^3 enter Rhode Island Sound from the West Passage and $0.273 \cdot 10^9$ m^3 from the East Passage. Thus, by this estimate, the water leaving the bay, and therefore the fraction of new water entering the bay, is 0.28 of the transport into Rhode Island Sound from the West Passage, and 0.18 of that entering Rhode Island Sound for the East Passage.

2. Another poorly known parameter is the average longshore drift in Rhode Island Sound past the mouth of the bay. Estimates vary considerably, but 5–10 cm s^{-1} would seem to be a reasonable value (Collins, 1974). If we consider the tidal volume to move as a slug in and out of the bay mouth, a crude estimate of the loss that would occur during the ebb may be made. During a 12 h tidal cycle starting at mean water, the leading edge of the block is beyond the boundary for a full 6 h. The trailing edge is out for only a moment at slack ebb. On the average then, the block will spend 3 h per cycle, or 6 h per day beyond the mouth of the bay, where it is

subject to the long-shore drift in the sound. The ratios of the drift rates of 5–10 $cm\,s^{-1}$ (1.08–2.16 km $6\,h^{-1}$) to the width of the passage mouths (2.8 km at the surface of the West Passage, 3.7 East Passage) indicate 0.61–0.23 of the West Passage flood and 0.71 to 0.42 of the East Passage flood tide is returning bay water.

3. Both previous "estimates" ignore the dominant aspect of estuarine circulation, two-layered opposing flow. Ed Levine (URI, personal communication) suggested a simple flushing calculation which addresses this problem. Consider a box model of two-layered flow:

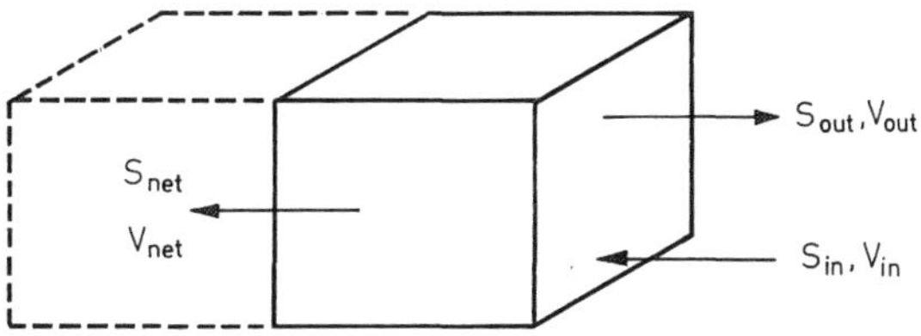

where S and V are the salinities and volumes of the incoming (in) and outgoing (out) water masses, and the net upbay transport (net). By conservation of mass it can be shown that the ratio of V_{in} to V_{out} *is a* function of only the salinities:

$$F = \frac{V_{in}}{V_{out}} = \frac{(S_{out} - S_{net})}{(S_{in} - S_{net})}.$$

Using the data of Hicks compiled by Hess (1974), this calculation can be made for three conditions of river flow. S_{in} and S_{out} are taken to be the observed bottom and surface salinities at the mouth of the passages, and S_{net} as the simple average of the surface and bottom salinity at the upbay boundaries of the large spatial elements (5 and 8, Fig. 1). These calculations suggest about 0.30 new water returning in the West Passage and 0.60 new water in the East Passage.

4. The model itself may be used to converge on the appropriate values of the Rhode Island Sound exchange parameters. Given observations of the salinity offshore and within the bay, the predicted salinity distribution should improve as the correct exchange factors are approached. A program was written which iteratively adjusted the West and East Passage exchange factors in order to achieve the best agreement with observed salinity data for the lower four elements of the bay using the criterion of minimizing the chi-square statistic, $\Sigma (x - \hat{x})^2$. The input data were taken from Hicks as compiled by Hess (1974, Table 7.1) for three river flow conditions. Although the zero flow condition is artificial considering the typical seasonal pattern (Fig. 13), the program converged strongly in all three cases producing these estimates for the exchanges:

River Flow $m^3/day \cdot 10^6$	West Passage factor	East Passage factor	Chi^2
0	0.80	0.90	0.0120
2.5	0.59	0.81	0.0002
5.0	0.52	0.67	0.0091
Average	0.64	0.79	

5. The four previous approaches provided estimates of the actual exchange of water at the mouth of the bay. In the model the primary interest is in the concentration of a material in the returning water. This is affected by the difference in concentration within and outside the bay as well as the physical exchange of water. By using observed nutrient levels instead of salinity as in approach 4, the model converges on some composite exchange–dilution factor. Again, the assumptions here are many and tenuous. During the fall of 1972, observations of nutrient levels roughly approached a plateau during a period of low planktonic activity (Figs. 4 and 5). By forcing the concentrations that were observed in the Providence River and allowing the model to equilibrate for different exchange–dilution factors, convergence toward the plateau concentrations in the lower bay was tested. With sewage and sediment fluxes from the model included, convergence occurred at the following values:

	West	East	Chi^2
NH_4^+	0.71	0.51	0.54
$PO_4^=$	0.86	0.75	0.45
$Si(OH)_3$	0.81	0.81	0.83
Average	0.79	0.69	

In summary, the various estimates seem to suggest the order of the exchange factors with remarkable consistency, considering the uncertainty in their assumptions (Table 6). The variability is greater for the East Passage, perhaps since the greater depths and volumes involved leave more room for error, and stronger two-layered circulation may be expected. Nevertheless, the appropriate range of values to use seems to be 0.6–0.8 and 0.7–0.9 for the West and East Passages respectively. That is, as a result of tidal exchange with Rhode Island Sound, concentrations in the waters returning to the lower bay after two tidal cycles may be expected to be diluted to between 60 and 90% of the value at the beginning of the day. The model allows for differing factors for each of the substances to be mixed in order to compensate for different offshore relative concentrations. Thus, phosphate, which remains at offshore levels about equal to those of the lower bay (Smayda, 1973a),

Table 6. Summary of Rhode Island Sound boundary exchange estimates – fraction of the ebb volume returning on the flood one day later

Method	West Passage factor WRIS	East Passage factor ERIS
1. Flushing time	0.72	0.82
2. Longshore drift	0.61–0.23	0.71–0.42
3. Box model	0.70	0.40
4. Salinity convergence	0.80–0.52	0.90–0.67
5. Nutrient convergence	0.79	0.61

may use exchange–dilution factors of near 1.0, while phytoplankton, which are relatively low in the offshore water entering the bay, may use factors based only on exchange of around 0.6 and 0.7.

3.5 River Flow

The three largest rivers entering Narragansett Bay are the Blackstone and Pawtuxet, which join as the Providence River entering spatial Element 1, and the Taunton, which enters Mt. Hope Bay. Historical flow data for these rivers, available through the U.S. Geological Survey (Surface Water Records for Mass., NH, RI and VT), have been compiled by Hess and White (1974). Hicks (1959) estimated that 72% of the total Taunton flow enters the bay under the Mt. Hope Bridge. From these facts, the following formulation for fresh-water input into the model was developed.

The combined flow is represented by a sinusoidal function reaching a maximum of $8.2 \cdot 10^6$ m^3/day with an amplitude of $4.1 \cdot 10^6$. During the late summer the flow is truncated so as never to fall below the $1.7 \cdot 10^6$ m^3/day minimum. The equation for this function is:

$$Q\ (\mathrm{m}^3/\mathrm{day}) = 4.1 \cdot 10^6 + 4.1 \cdot 10^6 * \cos[2\pi\ (\text{day-}60)/365] \qquad (Q \geqq 1.7 \cdot 10^6). \quad (6)$$

This total river flow is partitioned with 28% entering from Mt. Hope Bay (into Element 6) and the remainder into the Providence River (Element 1; Fig. 13).

The tidal mixing submodel TIDE, then, requires as daily input Taunton and Providence River flows, and concentrations of any number of substances in the

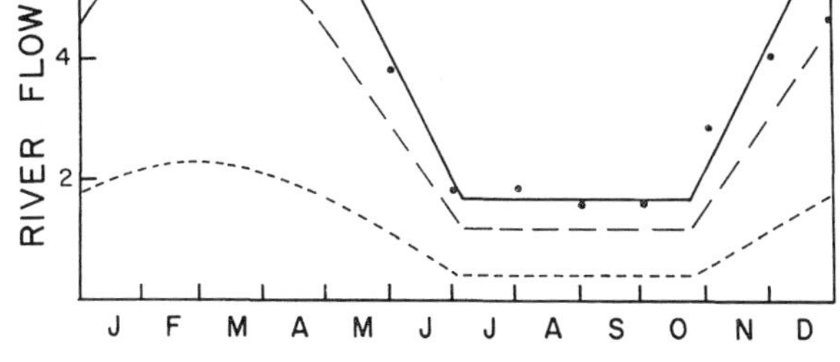

Fig. 13. Seasonal pattern of river flow in the Narragansett Bay model derived from data given by Hess and White (1974). *Upper curve*: total fresh-water input, followed by the flow of the Providence River into element 1 and the flow of the Taunton River into element 6

eight spatial elements. Tabular daily tidal ranges for one year at Newport are included within the program so that realistic day-to-day variation occurs. The routine simply returns to the central program new concentrations resulting from the stimulated mixing over two tidal cycles. While the scheme is extremely simplified, its consistency with other more sophisticated models suggests that the gross features of tidal circulation in Narragansett Bay are represented.

4. Phytoplankton

4.1 Temperature–Growth Relationship

The keystone of the phytoplankton formulation (Fig. 14) is the maximum growth rate, which is given as an exponential function of temperature. Since photosynthesis and consequent growth are physiological processes, this would appear to be an acceptable assumption, and a Q_{10} of about 2.0 might be expected. In fact, in a review of the subject, Eppley (1972) presents strong evidence for an upper physiological limit to phytoplankton growth in conditions where neither light nor nutrients were believed to be limiting (Fig. 15A). He suggests the equation:

$$\log_{10}\mu = 0.0275 \cdot \text{Temp} - 0.070$$

where: μ = specific growth rate, divisions per day

and Temp = temperature, °C.

We may rewrite the expression in exponential form of the base e:

$$\mu = 0.8511\, e^{0.0633 \cdot \text{Temp}}.$$

Both equations calculate divisions per day, which when used with the Base 2, provide an estimate of maximum population growth. For our purposes in the model, the rate in divisions per day ($\log_2$) is converted to the natural base ($\log_e$):

$$\text{GMAX} = 0.693 \cdot \mu$$

or

$$\text{GMAX} = 0.59\, e^{0.0633 \cdot \text{Temp}}\,(\text{day}^{-1}). \quad (7)$$

GMAX in this form is the instantaneous rate coefficient to be used in the eponential phytoplankton growth equation. It should be pointed out that the Q_{10} for this expression is 1.88. Further, recent work continuous to agree remarkably well with Eppley's formulation (Goldman and Carpenter, 1974), including observations on natural and cultured phytoplankton from Narragansett Bay (Smayda, 1973).

While the expression is believed to be quite sound for its application in the model, its use must be accompanied by an awareness of its limitations. As expected, the detailed picture becomes complicated, primarily due to specific responses of different species. Eppley (1972) points out that temperature optima frequently exist for laboratory studies of growth. These optimal rates may exceed the predictions of

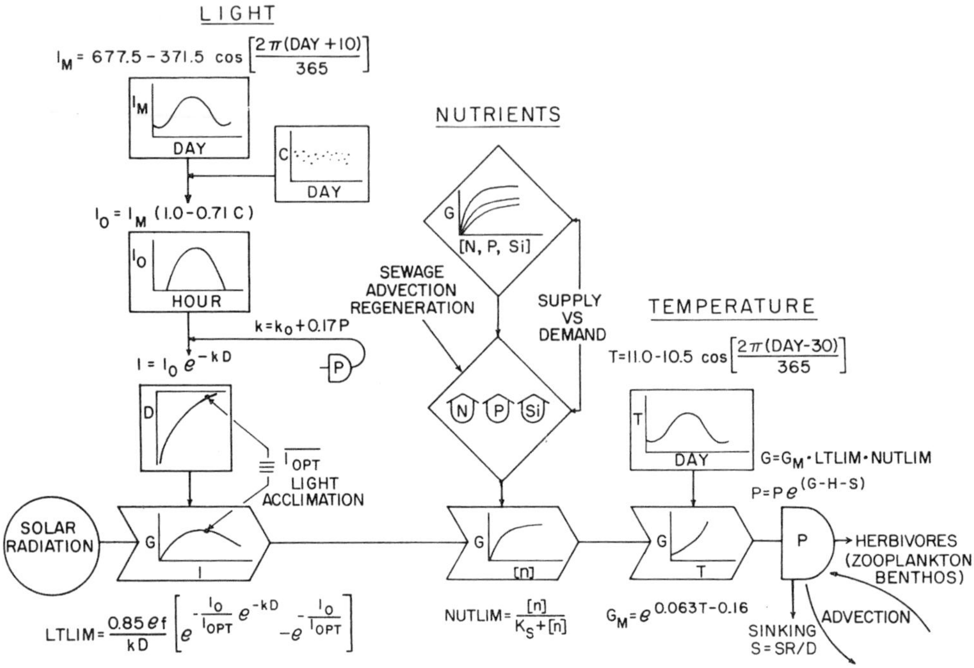

Fig. 14. Flow diagram for the phytoplankton compartment showing relationships among the major equations described in text and a graphical representation of their behavior

Eq. (7) (Durbin, 1974). Interestingly, however, in nature species are often most successful at conditions other than their laboratory optima (Smayda, 1969). This would apparently suggest favorable overall competitive advantage in conditions differing from the optimum for any single factor.

A second consideration is that Eppley's curve represents sustained growth rates, and it is likely that higher rates may be achieved for short periods under especially favorable conditions.

Another observation is that the apparently reasonable Q_{10} is lower than that suggested by a number of measurements in natural waters of about 2.3 (see Eppley, 1972). No clear explanation of this is forthcoming, although it may be related to the inadequacy of a general representation of many individual responses. In fact, perhaps the most serious limitation of Eq. (7) is that it obscures the most characteristic response of any single species (Fig. 15B). Thermal thresholds clearly exist for any organism, even though they may be adjusted by acclimation. Specifically, the typical response shows an increase with temperature to some upper limit, beyond which a rapid decline is observed (Eppley, 1972). For our initial purposes, the general response is a satisfactory representation since the species succession favored by seasonal temperature changes may be expected to result in the smoothed trend of Eq. (7). When this succession is considered explicitly, or if more detailed analysis of multispecies interactions were added to the model, such individual species patterns could be specified. A preliminary treatment of this type has been included in the model, allowing two "species-groups" with differing temperature responses to be simulated. This alternative formulation proposes two

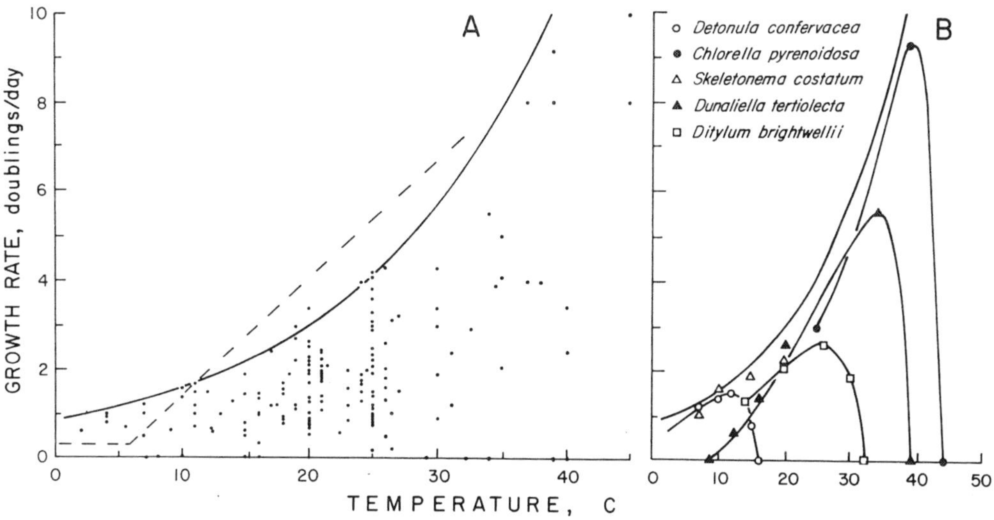

Fig. 15 A and B. Temperature response of phytoplankton growth rate. (A) Measurements of maximum specific growth rate in doublings per day for laboratory cultures largely in continuous light (Epply, 1972). *Solid line:* hypothetical maximum Eq. (7) used in most simulations. *Dashed line:* used in some runs of the model to represent a second, warm-water species-group. (B) Observations of specific growth rates for five species of unicellular algae demonstrating patterns of thermal optima which underlie the hypothetical maximum (Eppley, 1972)

linear equations which approximate Eq. (7) above and below the mean bay temperature of 11.5° C (Fig. 15A). While only such simple schemes have been tried so far, more complicated versions, such as translational and rotational thermal acclimation responses by different species (Goldman and Carpenter, 1974), could be included.

The data compiled by Eppley (1972) and expressed by the temperature-dependent equation represent realized growth per day. Thus, this formulation estimates net daytime phytoplankton productivity, implicitly including the respiratory losses. It is convenient to use this formulation because it obviates the additional calculation of respiration. However, nighttime losses are not included. While this omission introduces some error, it is especially desirable since the literature concerning phytoplankton respiration rates is quite uncertain (Yentsch, 1975). Even without the complexity of photorespiration, which may alter the rate during daylight hours (Morris and Beardall, 1975), estimates are difficult to generalize as a regular function of temperature. Further, observations in situ on natural populations of Narragansett Bay phytoplankton (Smayda, 1973) strongly confirm the adequacy of Eq. (7) in predicting an upper limit to realized daily growth.

4.2 Effects of Nutrients

The hyperbolic effect of nutrient concentration on the growth of microorganisms has long been recognized. Increasing concentration at low levels results in a commensurate increase in growth. At higher levels, the linear response is shifted,

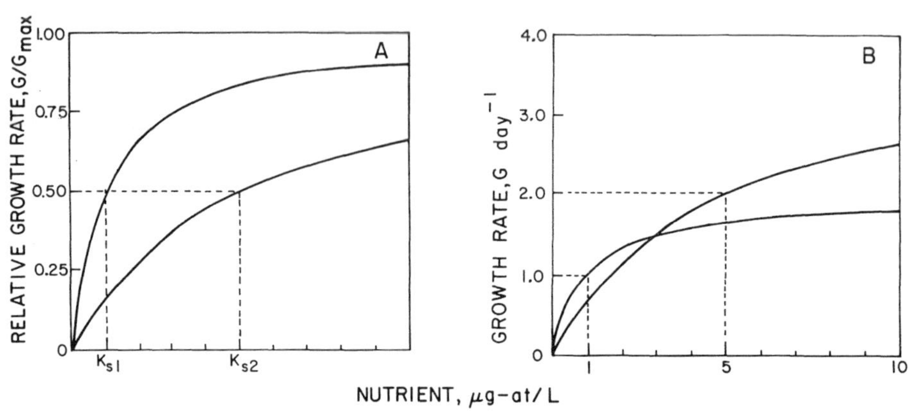

Fig. 16 A and B. Hyperbolic response of phytoplankton growth to a limiting nutrient. The half-saturation constants (K_s) are defined as the concentration at which growth is one-half the maximum. When growth is normalized to a maximum of 1.0, conclusions about the relative competitive advantage of species may be unwarranted. Species 1, with the lower K_s, appears totally dominant in the normalized representation (A). Consideration of the actual growth rates, however, reveals that species 2 grows faster at nutrient levels above 3 µg-at/l (B)

and the increment in growth diminishes for unit increases in the nutrient (Fig. 16A). Monod (1942) suggested an equation to express this relationship of diminishing returns:

$$G = G_{max}\left(\frac{N}{K_s + N}\right). \tag{8}$$

Thus, the predicted rate, G, is a function of the maximum (G_{max}), the ambient steady-state nutrient concentration (N), and a characteristic half saturation constant (K_s) defined as the concentration at which the rate is one-half the maximum (Fig. 16A). The effect of smaller values of K_s is to steepen the rate of ascent to G_{max}. The sense of the equation is intuitive—when N is much less than K_s, the fractional increase of the term in parentheses is almost linear, while as N approaches and exceeds K_s, the term approaches a maximum value of 1.0. Clearly, when $N = K_s$, the defined value of 0.5 is achieved.

The Monod formulation was suggested for substrate limitation of growth. A theoretical distinction may be made between the identical expression suggested by the biochemists Michaelis and Menten (1913) for the kinetics of enzymic reactions, since the limitations result at fundamentally different levels of organization. The extension of the Michaelis-Menten enzyme form to nutrient uptake or to growth is perhaps reasonable since these processes are the result of a combination of enzymic biochemical reactions. A stronger argument relates the processes in theory as special cases in the general phenomenon of what H. T. Odum refers to as double-flux reactions (Odum, 1972). The hyperbolic response is characteristic, although it may result from other functional units as well. For our purposes, the distinction is not critical, as the empirical agreement of the Monod expression is well established (Dugdale, 1967; Eppley and Coatsworth, 1968; Thomas and Dodson, 1968; Eppley and Strickland, 1968; Eppley and Thomas, 1969; Eppley et al., 1969; Hanton, 1969;

MacIsaac and Dugdale, 1969; Carpenter and Guillard, 1971; Fuhs et al., 1972; Paasche, 1973, 1973a).

Several theoretical constraints on the application of the Monod expression to dynamic populations have been frequently ignored. First, the degree of nutrient limitation is assumed to be independent of the standing stock of phytoplankton. That is, it is assumed that the uptake and growth of the algae does not affect the concentration of the nutrient. While it is theoretically sound that nutrient concentration is the property of the environment that is sensed and responded to by the organism, growth is not independent when changes in ambient nutrient levels result. This constraint may be met only in chemostat cultures or rare moments in nature, but numerous batch-culture experiments suggest that the theory holds reasonably well, even when this condition is not satisfied. In the model, it is important that the computations are made frequently enough to ensure that the assumption is not violated too seriously. In addition, a check may be made, which reduces the growth rate if the projected growth substantially alters the nutrient regime.

A second limitation to the use of Monod nutrient kinetics concerns the simplifying assumption of an "average" phytoplankton population. Models of this type generally consider only one such compartment for technical simplicity, although complex multispecies schemes are possible and necessary for the treatment of species competition and succession. The errors associated with representing multispecies populations by an average K_s choice have been investigated in a study using numerical modeling (Williams, 1973). While these errors may be significant in diverse communities with numbers equitably distributed among many species, the effect is reduced when one species dominates the composition or, as expected, if assumed kinetics of the separate species are similar. Both these assumptions may be better for neritic waters than for the open sea, but some misrepresentation necessarily occurs. For Narragansett Bay, the more critical question is probably not the averaging of diverse community properties, but rather the choice of a single value, thus ignoring the successional shifts in species dominance. There have been numerous measurements of K_s values for a great many species of marine phytoplankton, though most of these have been for nitrogen and most have been obtained for nutrient uptake rather than growth (e.g., Eppley and Thomas, 1969; Eppley et al., 1969; Carpenter and Guillard, 1971; McCarthy, 1972). There is considerably less information available on the uptake kinetics of phosphate (Hanton, 1969) or silicate (Paasche, 1973a; Mitchell-Innes, 1973). In later chapters we will discuss two approaches to this problem, including the general sensitivity of the model to choices of K_s, and a preliminary attempt to represent two species groups with differing nutrient kinetic properties.

The third inherent weakness of the simple Monod expression is the strict implication of a single limiting nutrient. Like the previous case, this is a limitation more in the application of the formulation than of the theory itself. The Monod equation satisfactorily represents the case when only one nutrient is limiting, and all other requirements are in excess. It is generally assumed that this approximation is similarly valid for the most limiting single nutrient, even if others are also quite scarce. Many models (DiToro et al., 1971; Dugdale and MacIsaac, 1971; Walsh and Dugdale, 1971) accept this and choose one nutrient as limiting, usually on the basis of the Monod calculation. Walsh (1975) compares light and nutrients as identical requirements, only one of which is allowed to reduce growth at any time. While the single limiting nutrient approach is taken in this model, multiple limitation is a possibility which should be considered. To date, no definitive study of such simultaneous limitation exists. Ketchum (1939) presents some data for nitrogen and phosphorus interaction which have been interpreted by DiToro et al. (1971) as supporting a multiplicative limitation effect. These authors however ignored their findings and chose to consider a single limiting nutrient in their model. Waite and Mitchell (1972) reported multifactoral experiments depicting the growth of *Ulva lactuca* with ammonia and phosphate interactions as a complex response surface. No simple interpretation of this surface seems possible, as it supports neither the single limiting factor nor multiplicative interaction view.

As a theoretical exercise, it is hard to decide a priori what formulation is most reasonable. An interaction scheme where limitation terms for all nutrients under consideration are multiplied together is not unreasonable, i.e.

$$G = G_{max} \frac{\mathrm{N}}{K_{s_N} + \mathrm{N}} \cdot \frac{\mathrm{P}}{K_{s_P} + \mathrm{P}} \cdot \frac{\mathrm{Si}}{K_{s_{Si}} + \mathrm{Si}} \cdots . \tag{9}$$

In fact, this equation reduces to Eq. (8) when all but one are in great excess. However, since the limitation terms of the equation never are exactly 1.0, multiplicative treatment has the disadvantage in practice of resulting in limitation being somewhat a function simply of how many nutrients are chosen for consideration. For example, the decision to include 10 required factors, all of which are only 0.95 limiting reduces the growth to 0.6, while a less-detailed treatment of only one factor would allow 0.95 of the maximum rate. Further, the problem is not readily resolved by natural observations, since the numerical effect of additional nutrient terms may be almost indistinguishable from a higher K_s value for the limiting nutrient, because concentrations of different nutrients tend to change together. Partly because of these numerical artifacts, but also since Liebig's law of the minimum continues to be invoked by most workers, the primary formulation in the model is to choose the single most limiting nutrient despite evidence for multiple interactions. The multiplicative method is easily compared however, and the results will be discussed later.

Because the maximum growth rate is determined independently according to Eq. (7), only the limitation term of Eq. (8) (in parentheses) is required. For the model, the equation has been normalized to approach unity by division of G_{max}:

$$\mathrm{NUTLIM} = \frac{G}{G_{max}} = \frac{N}{K_s + N} . \tag{10}$$

NUTLIM is a unitless fraction reflecting the degree of limitation by the nutrient, N.

The choice of the single limiting nutrient for the model is made by a comparison of these fractions for the three nutrients: total inorganic nitrogen, phosphate, and silicate. This approach avoids inconsistencies that may result from an alternative in which the limiting factor is chosen by a comparison of ambient nutrient ratios and phytoplankton elemental composition (Walsh, 1975). For example, if ambient nutrient concentrations are 6.0 μg-at/l nitrogen and 0.5 μg-at/l phosphorus, their ratios (12 : 1) when compared to the phytoplankton composition of 16 : 1, identify nitrogen as the limiting nutrient. However, kinetic calculations projecting uptake based on half-saturation constants for nitrogen of 1.5 and phosphorus of 0.25 (Walsh, 1975) suggest that phosphorus limitation will reduce growth to 67% of maximum versus only a reduction to 80% by nitrogen. Thus the nutrient identified as most limiting in the first scheme (nitrogen) is not the one least available for growth, and the projected uptake of phosphorus will exceed the physiological capacity indicated by the half-saturation constant.

Yentsch and Lee (1966) elegantly demonstrated the pitfalls of dealing exclusively with normalized curves of growth limitation such as Eq. (10). While their analysis concerned light acclimation and will be discussed further in that context, identical problems result when nutrient competition is being considered. In particular, for two or more species competing for the same nutrient, the unitless fraction term (NUTLIM) is not sufficient to evaluate which species has the competitive advantage. Both actual rate and limitation must be combined. In the example shown in Figure 16, the normalized curves (A) indicate species 1 with the lower K_s has the advantage at all nutrient concentrations. But when the relative maximum rates are considered (B), species 1 is only favored at low [N], with species 2 clearly dominant at all higher concentrations. In the basic model of one phytoplankton species-group, this is not a problem, but the formulation of two groups requires consideration of this factor when nutrients are scarce enough to force differential reductions in Monod-based growth estimates due to supplies insufficient to meet both demands.

The simple Monod kinetic expression, while currently the most commonly used, is not the only theory available for nutrient limitations of growth. More precisely, it has been expanded to account for evidence implicating intracellular nutrient storages. For example, both in cultures and in nature, growth is often observed to continue after apparent depletion of one or more nutrients from the environment. Further, the widely variable elemental composition of even single species demonstrates that nutrients are not always taken up in strict stochiometric proportions. The concept of luxury uptake (see Fogg, 1965) confers a strong advantage to a species that can take up and store nutrients in times of plenty to minimize direct competition as concentrations drop. Grenney et al. (1973) and Droop (1973) have presented theoretical formulations, supported by chemostat experiments, which postulate multiple extra- and intracellular nutrient pools that ultimately control growth in a Monod-like hyperbolic form. Such formulations provide valuable additional insight into multispecies interactions (Grenney et al., 1973) and luxury uptake under multiple nutrient interactions (Droop, 1973), but they require data on the size and turnover kinetics of the subcellular pools, which are only beginning to become available. Further, for a homogeneous species group without competition, such complex schemes can be expected to follow closely the patterns of simple Monod kinetics, differing perhaps by short phase shifts in the reciprocal patterns of phytoplankton–nutrient oscillations. The additional programming complexity was not considered worth the possible improvement in the basic model. However, the model does allow for a range of carbon : nutrient uptake ratios and a luxury kinetics scheme may be provided in order to evaluate an hypothesis. A simple hypothetical scheme has been prepared, which considers enhanced growth due to stored nutrients. The details of the scheme will be mentioned in a later section, with a discussion of its effects on the simulated phytoplankton dynamics. As more detailed information becomes available, more realistic schemes may be easily applied to the model as externally supplied subroutines.

4.3 Effects of Light

A large number of considerations are involved in estimating the extent to which light is restricting the ability of the phytoplankton to achieve the maximum temperature-dependent growth rate. The most fundamental of these is a characterization of the basic photosynthesis–light response. For this, the formulation of Steele (1962) was selected. This formulation involves designation of an optimum light level for photosynthesis. The optimum is held constant in most models, but because of the potential importance of light acclimation, an attempt has been made here to simulate variations in optimal light levels. The feedback on increased light extinction due to self-shading by the phytoplankton is also included. And finally, the effects of this combined formulation are evaluated throughout the daylight hours and through the depth of the water column to yield a single daily estimate of light limitation.

4.3.1 Theoretical Background

The derivation of a single expression for the effect of incident radiation on the daily growth of the entire phytoplankton population begins with an analysis of the instantaneous photosynthesis–light response. At the outset, the distinction must be made between patterns which characterize the instantaneous physiological response and those which represent a realized or integrated response over a relatively longer time period. Mathematically, it is possible to project the latter from the former, which is the approach chosen here and in many other models. It should also be possible, however, to work directly with empirical integrated responses over longer time periods, thus removing some potential artifacts and uncertainty. In fact, as will be shown, some differences in observed instantaneous response patterns may be explicated by theoretical analysis.

The instantaneous photosynthetic response has frequently been demonstrated experimentally. At low levels, photosynthesis increases roughly in proportion to available light until saturation begins to flatten the response. At higher levels, a region of inhibition occurs, although width of the plateau at saturation is apparently variable (Fig. 17). In many experiments, the saturation region appears to peak at a single optimum point, with significant inhibition observed with additional light (Ryther, 1956; Steele, 1962; Nixon et al., in preparation). The development from the early characterizations of a hyperbolic response has been reviewed repeatedly (Steele, 1962; Vollenweider, 1965; Fee, 1969, 1973, 1973a; Bannister,

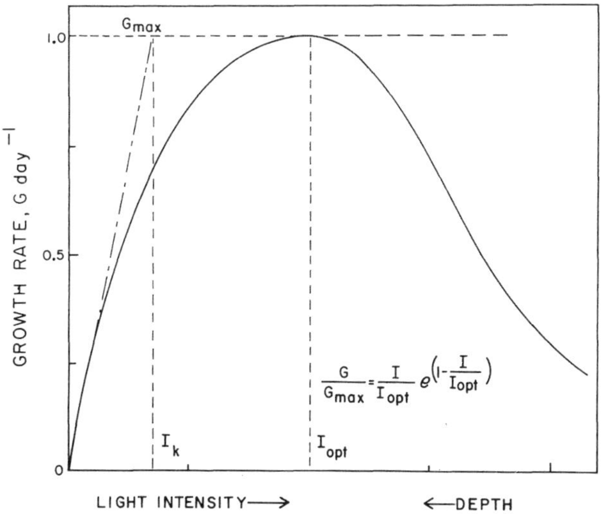

Fig. 17. Theoretical formulation for the instantaneous photosynthesis-light response of phytoplankton (Steele, 1962). The normalized Eq. (12) predicts growth relative to the maximum as a function of the ratio of the incident and the optimum light intensity, I/I_{opt}. Some experimental observations demonstrate a plateau with little or no high light inhibition. In such cases, another measure of light response may be used, with I_k defined as the intersection of the initial slope of the hyperbola with the maximum growth rate

1974, 1974a) and some rather complicated formulations have emerged in order to provide flexibility in curve-fitting the variable degree of high-level inhibition that is observed. Recent work has also produced theoretical analyses suggesting more fundamental expressions of the photosynthetic response to light quanta (Bannister, 1974). The equation proposed by Steele (1962) remains one of the more simple and remarkably versatile:

$$G = G_{max} \times \frac{I}{I_{opt}} e^{(1 - I/I_{opt})} \tag{11}$$

or

$$G/G_{max} = \frac{I}{I_{opt}} e^{(1 - I/I_{opt})}. \tag{12}$$

The degree to which the Steele equation approximates a wide range of photosynthesis–light observations is quite satisfactory (Ryther and Menzel, 1959; Steele, 1962, 1965; Nixon et al., in preparation) and has the additional advantages in the normalized form of requiring only one arbitrary parameter, with a conceptually meaningful interpretation.

In the original form [Eq. (11)], and when it is compared to experimental data with growth or production units, the Steele equation contains a term for the maximum growth, G_{max}, that is achieved, by definition, at the optimum light intensity, I_{opt}. In the normalized form, Eq. (12), the relative growth is expressed solely as a function of the I_{opt} parameter and the variable incident light, I. The normalized form is a unitless fraction, appropriate to the theoretical context of interacting factors used in our model. An additional value of the parameter I_{opt} is that it provides a mechanism in the model for considering the pertinent question of light acclimation by the phytoplankton, as discussed later.

Yentsch and Lee (1966) have pointed out the dangers in comparing populations or making inferences about the degree of light acclimation from experimental data presented in the normalized form. Apparent shifts in I_{opt} may be expected to result from a number of factors, only one of which is true sun/shade acclimation, and the normalized presentation of data obscures these distinctions. In the model, however, the logic is reversed; by postulating the acclimation, use of the normalized equation to express it is justified.

The support in the literature is widespread for this response pattern in general, and for the Steele formulation in particular, and the direct evidence for Narragansett Bay phytoplankton populations is also significant. As part of the Bay Systems Ecology program, numerous in situ vertical productivity profiles were taken. During the period Spring 1972–Fall 1973, 34 profiles were obtained for depth variations in productivity, as measured by oxygen titration of 300 ml light and dark BOD bottles (Nixon et al., in preparation).

To assist in the interpretation of these short term (3–4 h) incubations, smooth curves of the form of Eq. (11) were fitted to the observations using a least squares criterion. Extinction coefficients were measured with a submarine photometer during the incubations, allowing graphs of production vs. light. While the degree of surface inhibition was variable, the fit of the Steele equation was generally very good, in fact better than the fit of the more-complicated program provided by Fee (1969). Figure 18 presents a range of observations. The same curve-fitting has been applied to 24 h ^{14}C assimilation experiments, also with satisfactory results (G. Hitchcock, URI, personal communication), even though inhibition was frequently less apparent, as would be expected for the longer incubation.

Some concern has been voiced by H.T. Odum at the University of Florida, H. Seliger at Johns Hopkins University, and by others about the meaning of the widely observed high-light or "surface" inhibition of photosynthesis. Some feel that this inhibitory response is an experimental artifact which is not representative of the behavior of free-living algae. This artifact might be due to "bottle effects" such as abnormal buildup or depletion of chemicals, or to lack of turbulent mixing of the plankton throughout the light gradient in the water column. Inhibition observed in short-term experiments with dilute populations argues against bottle effects (our incubations were 3–4 h), and we designed an experiment to address the circulation question.

If the inhibition in surface bottles is due to unusually long residence in bright surface waters, a comparison of total production in a stationary array of bottles to that of an array rapidly moving through the water column should show significant differences. In our experiment, replicate arrays of 16 bottles each were suspended in situ for a 4 h incubation. At 5–10 min intervals, one array with the bottles equally spaced on a loop of rope was rotated so that all bottles frequently changed position. Each bottle of the cycling array moved throughout the water column, being exposed to inhibitory near-surface illumination for only 10–15 min at a time. Throughout the incubation, however, both arrays had the same number of bottles at each depth. Because each had a slightly different light history, variations were observed in the

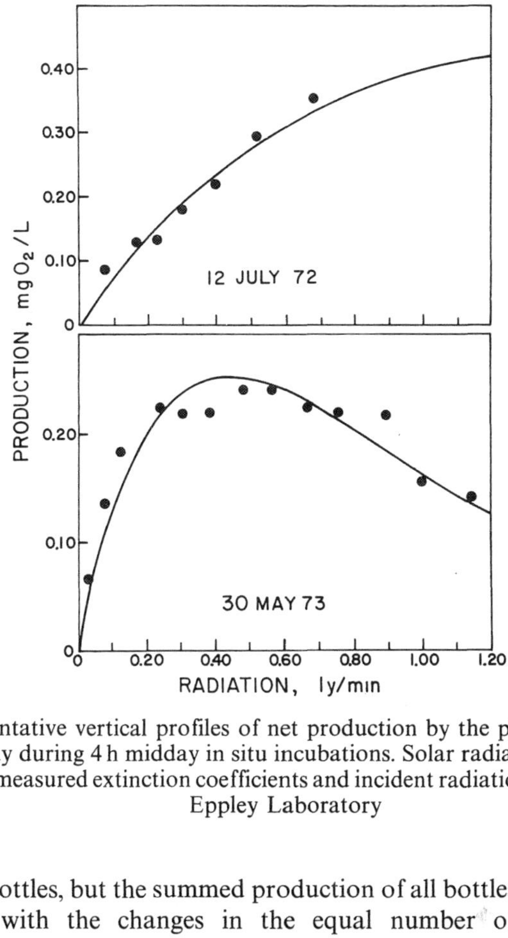

Fig. 18. Representative vertical profiles of net production by the plankton community in Narragansett Bay during 4 h midday in situ incubations. Solar radiation at each depth was calculated from measured extinction coefficients and incident radiation data provided by the Eppley Laboratory

cycling array bottles, but the summed production of all bottles (77.4 mg O_2) agreed quite closely with the changes in the equal number of stationary bottles (80.5 mg O_2). Calculating the increase that would have resulted if no surface inhibition occurred suggests a potential difference in the two arrays of 15%. This is sufficiently great to be distinguishable with the oxygen titration method used. While this evidence is preliminary, it provides additional support for the widespread observance of the high light inhibition in a variation of experimental situations, and the assumption of the proposed photosynthesis–light response.

After justification of the theoretical instantaneous response pattern as representative of natural phytoplankton populations, its employment in the model requires a transformation to account for the integrated effects of changing light regime both throughout the day and with depth of the water column (see Fig. 19). With some simplifying assumptions regarding light input, this integration may be performed analytically, and these approximations may be compared to a numerical integration of small depth and time increments and a more realistic light regime. The assumptions and difficulties in extrapolating daily production estimates from light–photosynthesis responses have been well reviewed by Vollenweider (1965). He concludes that the precise shape of the response curve may not be critically important in production estimates considering time and depth. Thus, the analysis

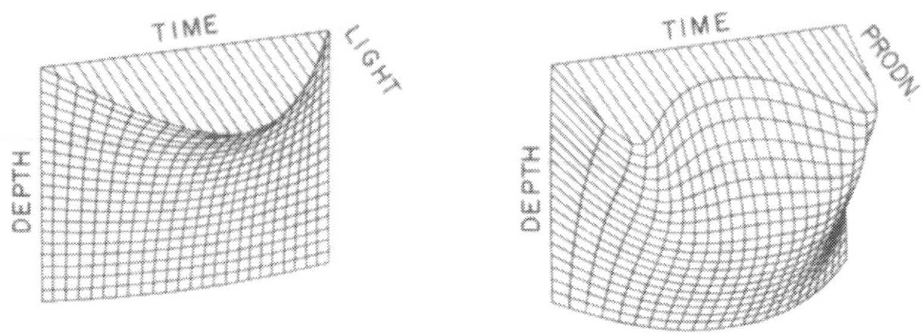

Fig. 19. Graphical representation of photosynthesis with depth throughout the daylight hours. Arbitrary conditions assumed incident radiation as a sinusoid, an I_{opt} of half daytime average, extinction coefficient $=0.25\,m^{-1}$, and a depth of 10 m

here using the Steele equation rather than more elaborate forms seems justified. Vollenweider continues, however, to emphasize that diurnal changes in the characteristic response, perhaps due to lags in nutrient availability, may be significant. Such considerations are beyond the temporal detail of this effort, and in the model, light responses are assumed to represent an average throughout each day. The weakness of the one-day timestep is probably restrictive here, even with the exact integration scheme used in the model. And the potential effects on the light response pattern during rapid nutrient recycling, which may occur–at least during the summer months in Narragansett Bay–are not adequately represented.

The use of the normalized Eq. (12) as a theoretical starting point minimizes these difficulties, which are critical in the interpretation of actual data. That is, in the model, the effect of a nonoptimum light regime is projected from the normalized relationship which avoids the apparent effect on I_{opt} of physiological state and nutrient or other stress. Such effects are believed to be mediated through changes in the dark reactions of photosynthesis and result in alterations of the original, unnormalized response pattern. While these stresses also effect the basic photosynthetic light reactions to some degree, the assumption that these are small in comparison to dark effects is commonly accepted (Steele, 1962; Yentsch and Lee, 1966). The formulation in this model makes this assumption, thus isolating factors related to light limitation from those of general physiological state in a way that is not possible when dealing with actual observations.

4.3.2 Time and Depth Integration of the Steele Equation

The curved pattern of diurnal irradiance in the simplest case of a cloudless sky makes analytical integration of even the relatively simple Steele equation impossible. Steele (1962) used an assumption of a linear increase from dawn to midday and subsequent decrease to dusk allowing a time-averaged production estimate in terms of the average light, $\bar{I}$:

$$\frac{\bar{G}}{G_{max}} = \frac{e \cdot I_{opt}}{2\bar{I}} \left[1 - \left(1 + \frac{2\bar{I}}{I_{opt}} \right) e^{-2\bar{I}/I_{opt}} \right]. \tag{13}$$

The depth integral of Eq. (13) under the exponential extinction of light becomes formidable and Steele does not present a solution. This formulation was not used in the model.

Another simplifying assumption invoked by DiToro et al. (1971) permits analytical solution of the depth–time integral, but with substantial loss of biological validity. If the incident radiation is taken to be constant through the entire photoperiod, i.e., $I = \bar{I}$ at all times, the exact double integral is (DiToro et al., 1971):

$$\frac{\bar{G}}{G_{max}} = \frac{e \cdot f}{k \cdot z}\left(e^{-\frac{\bar{I}}{I_{opt}} e^{-kz}} - e^{-\frac{\bar{I}}{I_{opt}}}\right), \tag{14}$$

where: e = the base of natural log,
k = extinction coefficient (m^{-1}),
z = the depth (m),
f = the photoperiod as a fraction (e.g., 0.5 at the equinoxes),
$\bar{I}, I_{opt}$ = incident average and optimal light (eg., ly/day).

A brief comparison of the equations at this point is helpful. Equation (12) is fundamentally different from the other two because it alone represents the idealized physiological response at any instant. Equation (12) represents the instantaneous rate in comparison to the integrated total production of Eqs. (13) and (14), i.e., mgC mgC^{-1} day^{-1} (rate) vs. mgC mgC^{-1} (in one day). It is immediately obvious from Figure 20 that the consideration of time and depth variations in light significantly dampens the basic pattern. It is interesting to note that Eq. (14) approximates a simple hyperbola, though a direct comparison reveals that no simple rectangular hyperbola of the Monod form agrees precisely. It is also interesting that in general, longer in situ incubations frequently result in such hyperbolic plateaus (Smayda and Hitchcock, personal communication; Talling 1971; see Fig. 21).

The plateau of Eq. (14) eventually begins to fall, demonstrating inhibition, but only when a finite depth is assumed. In meaningful terms, the theory predicts that increased surface radiation will result in increased total net production for the water column as long as the depth of optimal light (and therefore maximum production) remains above the bottom, or thermocline. In experimental situations, the effective "depth of the water column" is the height of the sample bottle, and inhibition might be expected even in in situ incubations if $I_{opt} \ll I$. This is also supported by the 24 h ^{14}C incubations of Smayda and Hitchcock, where marked inhibition was observed in an unusually dark season. This analysis, then, provides a general explanation for the range of photosynthesis–light observations demonstrating degrees of high-light inhibition from distinct, single-point optima, through simple hyperbolae, to broad optimal plateaus with some inhibition only at very high levels. The contention that surface inhibition is an experimental artifact may be resolved by the assertion that for the water column as a whole, surface inhibition rarely results in decreased net production with increased incident radiation, except in very shallow waters. The use of a Monod expression for light limitation in a diurnally and vertically varying grid array (see Walsh, 1975) is apparently less satisfactory than such an assumption would be for a vertically integrated production estimate.

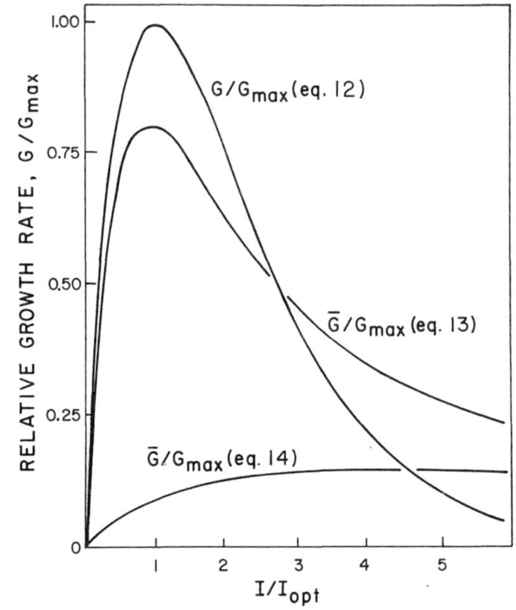

Fig. 20. Comparison of theoretical photosynthesis–light equations. *Upper curve:* instantaneous growth response for the basic physiological Eq. (12); in this case instantaneous growth rate as a function of irradiance is predicted. *Middle curve:* average daily growth at one depth integrated throughout the day assuming light increases linearly from dawn to midday and decreases linearly to sunset [Steele, 1962; Eq. (13)]. *Bottom curve:* integral through both time and depth, producing a 24 h average estimate assuming a square-wave average light input [DiToro et al., 1971; Eq. (14)]. Calculations were made with an extinction coefficient of $1.0\ m^{-1}$ and a depth of 10 m

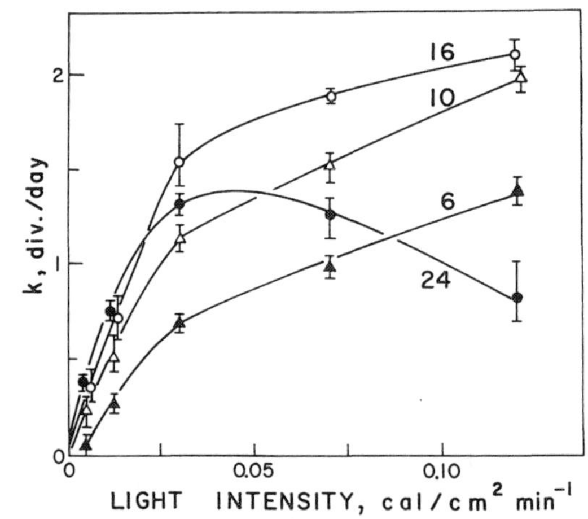

Fig. 21. Experimental observations on the effect of photoperiod on specific growth rate (from Paasche, 1968). Under continuous illumination (numbers indicate hours of light per day), maximum growth was depressed, and high-light inhibition resulted. In contrast, the hyperbolic pattern is at least partly a consequence of integrating the physiological photosynthesis response over a light–dark cycle (see text)

A final theoretical consideration concerning the DiToro double integral of Steele's equation is the weakness of the constant light assumption. The power of the Steele equation to represent surface inhibition is substantially diminished when midday brightness is lost in a daily average. To evaluate this error, a program was written to numerically integrate Eq. (12) over 0.25 m depth intervals every 10 min during the day. Radiation was assumed to follow a half-sine wave peaking at noon. The comparison of this result with the square-wave estimate [Eq. (14)] is presented in Figure 22A and B. The discrepancy is nonlinear and is increasingly severe for clear waters and high degrees of inhibition. While a curve could be fitted to provide

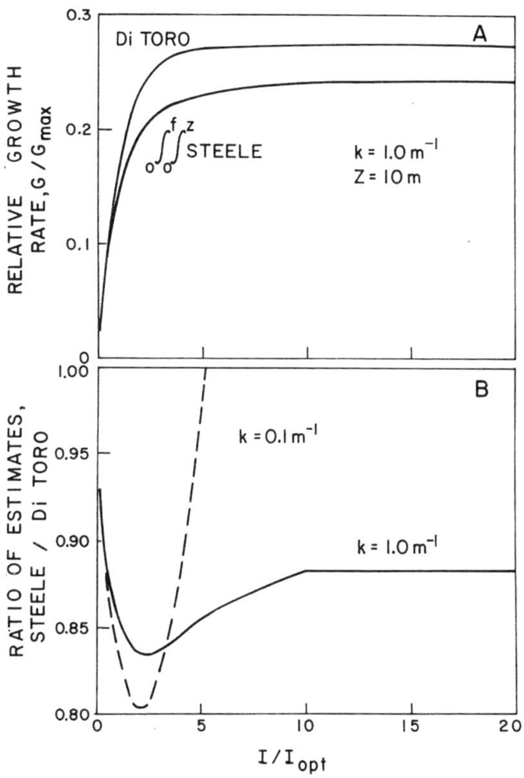

Fig. 22 A and B. Graphical demonstration of the error introduced into the double integral over time and depth of the photosynthesis-light equation of Steele (1962). The equation of DiToro et al. [1971; Eq. (14)] assumes a constant, average irradiance throughout the photoperiod, thus routinely overestimating production by the elimination of midday surface inhibition. Relative daytime growth from the DiToro equation has been compared with a numerical integration of the Steele equation assuming a more realistic sinusoidal pattern of irradiance (A). The ratio of the two estimates is also presented (B). For normal ranges of I/I_{opt} in Narragansett Bay, a constant correction factor of 0.85 was used with the DiToro equation. With appropriately low light attenuation, surface inhibition may extend to the bottom, resulting in a decline in both growth predictions. Ratios greater than 1.0 result in extreme cases ($k=0.1\ m^{-1}$ and $I \gg I_{opt}$) when the DiToro equation shows minimal growth throughout the day, while the sine integral predicts growth near dawn and dusk when light is optimal in the water column (B, *dashed line*)

a correction factor depending on I/I_{opt}, such detail is not justified and a constant overestimate of 15% by Eq. (14) has been assumed representative of the range of $I/I_{opt} = 1$–4.

The resulting expression used in the model estimates the accumulated effects of nonoptimum light throughout the water column during a 24 h day.

$$\mathrm{LTLIM} = 0.85 \cdot \mathrm{e} \cdot f\left(\mathrm{e}^{-(I/I_{opt})\mathrm{e}^{-kz}} - \mathrm{e}^{-I/I_{opt}}\right)/kz, \tag{15}$$

where the terms are the same as in Eq. (14). LTLIM is a unitless fraction to be multiplied times the daily production that would otherwise be achieved in continuous optimum illumination.

4.3.3 Light Acclimation and the Selection of I_{opt}

The foundation of the light formulation presented above is the parameter I_{opt}, the optimum illumination for photosynthesis of the phytoplankton population. The arguments for the existence of such a light level have already been presented here and elsewhere (Steele, 1962; Vollenweider, 1965; Fee, 1973, 1973a; Bannister, 1974; Goldman and Carpenter, 1974). The assumption of a single value to represent the light response of a mixed population seems justified, since numerous observations on natural communities demonstrate optima as clearly as unialgal cultures. Further support might follow from a line of reasoning parallel to that offered in justification of single nutrient half-saturation constants. But the selection of a constant I_{opt} value is more tenuous, despite its use in other models.

Significant changes in the optimum light response over short spans of time and space have been widely recognized for many years. Ryther and Menzel (1959) identified sun and shade acclimated populations resident in different depth layers of the Sargasso Sea. Numerous investigations have documented variations within and between species in the laboratory (Jitts et al., 1964; McAllister et al., 1964; Smayda, 1969; Ignatiades and Smayda, 1970), as well as in many mixed natural populations (Nixon et al., in preparation; Smayda and Hitchcock, personal communication). Yentsch and Lee (1966) demonstrated experimentally, and explained theoretically, alterations in the apparent initial light response characterized by a parameter, I_k, defined as the intersection of the initial slope of the photosynthesis–light response with the maximum production rate (Fig. 17). Its use was suggested to identify and compare the light reaction capacity based on hyperbolic patterns that demonstrated no clear-cut optimum. With the Steele equation, the initial slope can be shown by calculus to be I_{opt}/e, thus allowing direct comparison of these related parameters. Changes in I_k or I_{opt} may be due to a variety of external factors, only some of which actually reflect changes in the light reactions of photosynthesis (Yentsch and Lee, 1966). Under some experimental and natural conditions, however, these partial effects can be distinguished, and it is now well established that the photosynthetic capacity reflects to some degree the past light history of the species or community. The exact mechanisms involved in light acclimation are not well known, though it is generally assumed that changes in the chlorophyll content are closely involved. A great deal of work attempting to discern meaningful patterns in assimilation number (ratio of the rate of carbon fixation to chlorophyll content) with light and other factors is underway.

Because of the prominent occurrence of variation in light response patterns and the active research being done on the mechanisms and importance of light acclimation, it was desirable to include this detail in the model. However, a simplified approach—empirical rather than mechanistic—was chosen initially in order to simply allow a general assessment of potential importance. As observational data lead to a better understanding, or at least to firmer mechanistic hypotheses, more advanced formulations may be evaluated.

The basic assumption is that the optimum light for growth, I_{opt}, tracks the previous light history to which the algae have been exposed. Interestingly, no clear seasonal pattern or correlation with prior insolation was apparent in I_{opt} values from our depth productivity profiles (Nixon et al., in preparation). But similar observations using 24 h ^{14}C incubations revealed a tendency for I_k to correlate with light history during certain times of the year (Hitchcock and Smayda, 1977). Experimental data for *Chlorella* (Steemann-Nielsen et al., 1962) first suggested an approximate rate for acclimation (Fig. 23) of 2 or 3 days, with the most rapid change during the first day. Further work (Steemann-Nielsen and Park, 1964) continues to suggest this time frame for such changes in the response pattern of cultures and communities.

Another consideration in the specification of a changing I_{opt} is the light level to which acclimation is coupled. Our vertical productivity profiles suggested that fully acclimated natural bay populations frequently demonstrate maximum production at a depth of about 1 m. Over a wide range locations and seasons, the tendency was remarkably consistent. For extinction coefficients observed throughout the bay, the 1 m depth represents from 40–70% of I_{sfc} ($\bar{x}=51\%$). Considering the ratio $I_{opt}:I_{surface}$, Steele reports a number of observations which agree with his assumption that $I_{opt}=50\%\ I_{sfc}$ (see Steele, 1962: Steele and Baird, 1961, 50–100%; Steemann-Nielsen and Jensen, 1957, 33–50%; Sorokin et al., 1959, 30–50%), and the

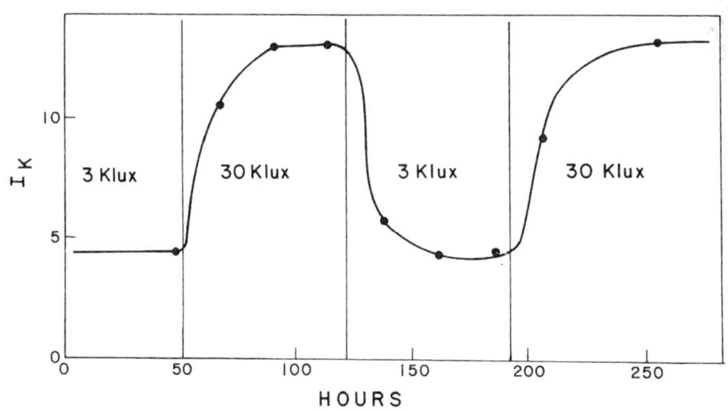

Fig. 23. The time-course of light acclimation by *Chlorella vulgaris* to changing intensity (from Steemann-Nielsen et al., 1962). I_k relates to the photosynthetic response at low light levels (Fig. 17). Based on these data, the model calculates the light of optimum growth I_{opt} as a weighted moving average of the previous three days insolation [Eq. (16)]

previously mentioned ^{14}C incubations in Narragansett Bay by Smayda and Hitchcock have hinted at a trend which also supports this estimate.

Based on these considerations, the model determines I_{opt} as a weighted moving average of the light intensity at the 1 m depth for the previous three days. The weighting factors were chosen to approximate the observations of Steemann-Nielsen et al. (1962):

$$I_{opt} = 0.7I'_1 + 0.2I'_2 + 0.1I'_3, \tag{16}$$

where I'_j the average light level at 1 m, j days earlier.

Under a constant light regime, the population will be 70% acclimated after one day, 90% after two days and 100% on the third day. Under the scheme of stochastic cloudiness in the model (Chap. 3), continuous acclimation occurs which will, on the average, reflect a seasonal trend of mean insolation.

The capacity for acclimation is clearly not unlimited—there certainly exist levels above and below which successful acclimation is not possible. In the model, omission of this consideration would mean, for example, that low winter levels of radiation have little limiting effect on growth. But work in Long Island Sound (Riley, 1967) and recently in Narragansett Bay (Hitchcock and Smayda, 1977; Nixon et al., in press) suggest that lower light thresholds exist which may exert important controlling influences on phytoplankton. Thus, the model includes the parameter LOIOPT below which the acclimation of Eq. (16) cannot occur. Riley (1967) observed that an average insolation in the water column (or mixed layer) of 40 ly/day appeared necessary for the onset of rapid growth in Long Island Sound. This numerical value was chosen for routine use in the model, and the effect of this as well as the choice of I_{opt} will be discussed later.

Throughout the light acclimation section, the distinction between instantaneous rates and insolation (integrated totals) must be made. Laboratory or short-term field observations attempt to assess the instantaneous rates of the photosynthesis–light response. That is, the maximum production and the optimum light are rates achieved for a short time—P_{max}(mgC mgC^{-1} day^{-1}) at an irradiance I_{opt}(ly day^{-1}). Longterm observations inherently integrate and average these rates resulting in fundamentally different values—$\bar{P}_{max}$(mgC mgC^{-1} in 1 day) at an insolation $\bar{I}_{opt}$ (ly received in 1 day). The distinction is one of degrees, since even short 4 h incubations represent an average, but the assumption that an instantaneous rate is being measured is more nearly satisfied. While this distinction may appear obvious, it is emphasized here because it appears to have been ignored in some applications of experimental observations to model formulations (see DiToro et al., 1971). Thus, it is inappropriate to use an I_{opt} based on instantaneous rate estimates from continuous illumination cultures in a formulation of ambient light regimes based on insolation measurements over the day. At the very least, an experimentally determined I_{opt} should more properly be compared with some estimate of the irradiance. For example, an insolation of 200 ly received over 24 h can be interpreted as suggesting a daytime average irradiance $\bar{I} = 400$ ly day^{-1} for 12 h of light, while during the mid-day hours, rates of $I = 600 - 800$ ly day^{-1} may occur.

An additional advantage of the variable I_{opt} formulation chosen here is that it avoids the problem of units discussed in the previous paragraph. Observations on instantaneous rates are fundamental to the scheme (the three-day time course of acclimation and the empirical 1 m optimum), but the model I_{opt} is calculated in whatever light units are given (here, insolation, ly day^{-1}) and is therefore necessarily consistent.

4.3.4 Self-Shading and Extinction Coefficient

As phytoplankton standing stock increases, the absorption of light by the cells themselves reduces the energy available to the population lower in the water column (Lorenzen, 1972). This self-shading is perhaps the most direct mode of the feedback control of the algae on their own growth, since nutrient availability involves other parts of the system. Self-shading is conceptually intuitive, and can be easily modeled by increasing the extinction coefficient of the water in proportion to plant biomass, so it is often included in modeling efforts (DiToro et al., 1971; Walsh, 1975). Formulations of the effect are based on empirical regressions relating these two variables, such as the equation of Riley (1956):

$$k = k_0 + 0.054\,C^{2/3} + 0.0088\,C\,, \tag{17}$$

where: k_0 = extinction of water with no chlorophyll (0.04 m^{-1}),
C = Chlorophyll *a* concentration (mg m^{-3}),
k = total apparent extinction coefficient (m^{-1}).

Application of this relationship to pelagic systems has been successful—indeed, Riley's fitted data spanned three orders of magnitude—and detailed discussion of its validity and interpretation continue (see Bannister, 1974, 1975; Riley, 1975). More simplified versions have been suggested which essentially group the pigment-dependent terms of Eq. (17) into a linear term:

$$k = k_0 + k_c \cdot C\,.$$

Presumably these curves approximate portions of Riley's nonlinear equation for appropriate ranges of biomass (see DiToro et al., 1971). But designation of a range for estuaries where plankton may span orders of magnitude is tenuous, and use of the original equation is advisable.

The intercept of Riley's equation (k_0) represents the extinction of water due to all nonchlorophyll related substances. In estuarine waters this intercept is undoubtedly higher than Riley's 0.04 m^{-1}, and is extremely variable due to large fluctuations in particulate load and general water quality. Schenck and Davis (1972) demonstrated that extinction coefficients measured in Narragansett Bay were at least as variable diurnally with tidal influences as seasonally due to plankton patterns. The only discernible pattern was a clear gradient from the Providence River to the mouth of the bay, paralleling the expected sediment load distribution.

Table 7. Estimates of the extinction coefficient, k, and the non-chlorophyll component, k_0, for 8 spatial regions of Narragansett Bay. k_0 was back-calculated from the equation of Riley (1956) and the data of Schenck and Davis (1972) with chlorophyll data for the same sampling period in one location (A), and for baywide chlorophyll data during another year (B)

Element	Observed[a] k m^{-1}	Estimated k_0	
		(A) m^{-1}	(B) m^{-1}
1	0.92	0.49	0.57
2	0.76	0.41	0.39
3	0.81	0.43	0.43
4	0.67	0.36	0.34
5	0.58	0.31	0.37
6	0.71	0.38	0.43
7	0.54	0.29	0.31
8	0.35	0.19	0.19

[a] Schenck and Davis (1972).

To estimate the nonchlorophyll-related component for the bay, the mean observed extinction coefficient for the lower West Passage of the bay was combined with an average chlorophyll estimate for the same time and place. Nine samples during July and August 1971 averaged 6.4 μg chlorophyll/l (±3.6 S.D.; Hitchcock and Smayda, personal communication), and Schenck and Davis (1972) report an extinction coefficient of $0.52\,m^{-1}$ (±0.13 S.D.). Back-calculating with Riley's equation gives a value for k_0 of $0.28\,m^{-1}$, with a range from 0.27 (minus one standard deviation for both k and Chl) to 0.31 (both plus one S.D.). Using the observed pattern of k throughout the bay (Schenck and Davis, 1972) to prorate this single value produces baywide estimates for k_0 (Table 7, Column A). In an alternate method, all the chlorophyll data for the summer months of the 1972–1973 sampling program (Nixon et al., in preparation) were paired with observed mean values of k for the appropriate bay region (Schenck and Davis, 1972) resulting in some 335 calculated k_0 estimates. The average of these approximations for the eight bay elements (Table 7, Column B) compares quite favorably with the estimate based on the chlorophyll observations for the same year as the transparency study.

4.3.5 Sinking Rate

The importance of physical settling of phytoplankton as a sink in Narragansett Bay is uncertain. Numerous measurements (Smayda, 1970) have estimated the sinking rate of diatoms to be between 0.25–2 m daily. Such rates are significant in the ocean, or in large bodies of water characterized by sluggish currents or incomplete mixing. The predominance of diatoms, especially during the colder months, might suggest that sinking is important in Narragansett Bay. However, a number of considerations argue the opposite.

Tidal mixing throughout the bay is extensive, resulting in current speeds of a knot or more. With the irregular basin geometry, relatively shallow depth, active wind field and low fresh water input, a well-mixed water column is the rule (Hicks, 1959). Stratification occurs in the Providence River, especially during the summer, but the well-mixed assumption is generally justified. This turbulence would be expected to minimize, or even negate the sinking tendency of diatoms. This conclusion is supported by analyses of sediment trap data and bottom sediment samples for Narragansett Bay (Oviatt and Nixon, 1975). Organic matter present in resuspended particulate material was not correlated with phytoplankton in the overlying water, even when short time lags were employed. Thus it would appear that sinking of diatoms may not be of major importance as a loss term.

In the interest of generality, however, and to enable testing of sinking rates as variable parameters in the model system, this phenomenon was included in the phytoplankton compartment. A sinking rate (SNKRAT) in m day^{-1} removes a fraction of the population daily in proportion to the local depth of the spatial element. Loss of phytoplankton due to sinking (SINK), and the resultant contributions of nutrients (SINKN, SINKP, SINKSI) in proper ratios, are thus lost to the sediments, and added to the net nutrient balance of the benthos (SEDN, SEDP, SEDSI; see Chap. 6). Presently the model only accepts a sinking rate for the first species group. The original motivation was that the second group would usually represent a flagellate-dominated population, where no sinking rate would be appropriate. In most simulations to date, however, SNKRAT has been set to zero with the assumption of mixing adequate to compensate for algal settling. The sinking formulation has been used primarily to suggest whether nutrients from diatom settling in addition to copepod fecal pellet contributions and benthic grazing is sufficient to balance the empirical sediment fluxes in the model.

4.4 Two Phytoplankton Species Groups

While the initial thrust of this model included only a single phytoplankton compartment, the interest and potential importance of seasonal shifts in certain community properties prompted a more-detailed formulation. In most cases these additions do not involve any fundamentally different approaches other than those already discussed. It should be emphasized that in these two compartments the formulations also attempt only to represent average responses of communities of species with generally similar characteristics. The difficulties in generalizing from single species remain, and detailed interspecific interactions are only crudely represented. While seasonal dominance due to temperature or nutrient preferences were the goal of this addition, it is perhaps equally useful in the general evaluation of these competitive attributes without specific regard to seasonal succession.

Presently the model provides two temperature formulations that may be applied to the two species-group system. Form 1 assumes the same temperature response for both groups as has been presented earlier for the single phytoplankton compartment Eq. (7). This allows analysis of nutrient effects alone. Form 2 depicts linear approximations of the basic Eppley curve which favor the second phytoplankton group for temperatures above the bay average of 11.5° C (Fig. 15A). The low-temperature rate is truncated at 0.25, and Group 1 retains the basic Eppley response. Again, while any single species tested acutely for temperature effects on growth might be expected to demonstrate a rapid decline above a critical point, seasonal community adaptation and acclimation would mask these responses within a natural temperature range. Critical maxima are not represented here for this reason, and effects of extreme temperature regimes cannot be tested at present. Such modifications are easily made, however, should this aspect need consideration.

Differing light responses have not been formulated for the two groups. The rationale for this is that short-term acclimation to seasonal insolation has already been included. It is assumed that the entire plant community undergoes this acclimation to about the same degree.

Another potential consideration which has not been included is differential grazing by the herbivores. Selection coefficients designating a feeding bias as some function of particle size or relative abundance (Berman and Richman, 1974; D. Heinle, personal communication) would be intriguing and straightforward to include as additional data become available.

The final factor of competitive interaction in the two-group system is different nutrient kinetics. The model accepts unequal half-saturation constants and carbon-nutrient ratios for each of the three nutrients (N, P, Si). This refinement weakens the simple limiting nutrient assumptions used in the single compartment version since relative nutrient limitation may no longer be considered independent of the projected growth (uptake) rates. Competitive advantage and relative nutrient uptake may depend not only on the ambient concentrations as suggested by Monod-type calculations, but also on the biomasses of the two species-groups. In earlier versions of the model, when the full one-day time-step was used, the chance that the combined demand would exceed the available supply was guarded against. The programming complexities required to anticipate all combinations of different nutrients limiting two groups with unequal growth rates and nutrient requirements are formidable, though not insurmountable. With the predictor–corrector integration scheme of the present version, however, these precautions have proven unnecessary (Chap. 7).

In the computer program, the statements for each calculation are repeated for both of the two species groups, P1MGC and P2MGC. For simplicity, the equations of the model are represented here in general form, but are essentially the same as the actual FORTRAN coding of the model:

$$\mathrm{GMAX} = 0.59 \cdot e^{(0.0633 \cdot \mathrm{TEMP})} \qquad (\mathrm{day}^{-1})$$

$$\mathrm{CHL} = \mathrm{PMGC} / \mathrm{C{:}CHL} \qquad (\mu g/l)$$

$$\mathrm{K} = \mathrm{K0} + 0.0088 \cdot \mathrm{CHL} + 0.054 \cdot \mathrm{CHL}^{(0.6667)} \qquad (\mathrm{m}^{-1})$$

$$\mathrm{IOPT} = 0.7 \cdot \mathrm{I1} + 0.2 \cdot \mathrm{I2} + 0.1 \cdot \mathrm{I3} \qquad (\mathrm{ly/day})$$

$$\mathrm{IF}\ (\mathrm{IOPT} < \mathrm{LOIOPT})\ \mathrm{IOPT} = \mathrm{LOIOPT}$$

$$\mathrm{TERM1} = \mathrm{DAYRAD} \cdot 0.9/\mathrm{IOPT}$$

$$\mathrm{TERM2} = \mathrm{TERM1} \cdot e^{(-\mathrm{K} \cdot \mathrm{DEPTH})}$$

$$\begin{aligned}
\text{LTLIM} &= 2.72 \cdot \text{PHOTPD} \cdot (e^{(-\text{TERM2})} - e^{(-\text{TERM1})}) \cdot 0.85/(\text{K} \cdot \text{DEPTH}) \\
\text{NTOT} &= \text{NH4} + \text{NO2NO3} && (\mu\text{g-at/l}) \\
\text{NLIM} &= \text{NTOT}/(\text{KN} + \text{NTOT}) \\
\text{PLIM} &= \text{PO4}/(\text{KP} + \text{PO4}) \\
\text{SILIM} &= \text{SI}/(\text{KSI} + \text{SI}) \\
\text{MXLIM} &= \text{AMIN1 (NLIM, PLIM, SILIM)} \\
\text{GP} &= \text{GMAX} \cdot \text{LTLIM} \cdot \text{MXLIM} && (\text{day}^{-1}) \\
\text{SINK} &= (\text{SNKRAT/DEPTH}) \cdot \text{PMGC} && (\text{mgC/l}) \\
\text{SINKN} &= \text{SINK} \cdot \text{N} : C_p && (\mu\text{g-at/l}) \\
\text{SINKP} &= \text{SINK} \cdot \text{P} : C_p && (\mu\text{g-at/l}) \\
\text{SINKSI} &= \text{SINK} \cdot \text{SI} : C_p && (\mu\text{g-at/l})
\end{aligned}$$

where:

PMGC = phytoplankton biomass, mgC/l (in the model P1MGC + P2MGC),

I1, I2, I3 = average insolation at 1 m one, two and three days ago [see Eq. (16)],

TERM1, TERM2 = intermediate values for the calculation of LTLIM,

DAYRAD = insolation [see Eq. (5)], reduced by 0.9 to correct for albedo,

KN, KP, KSI = nutrient half-saturation constants,

MXLIM = most limiting nutrient term, determined by the function AMIN1,

and all other variables are as defined previously in text.

5. Zooplankton

5.1 General Background

While the conceptual framework of the zooplankton compartment includes a number of commonly accepted processes, the representation of interactions in the formulation is perhaps the most complex to be found in the model. For example, conservation of mass in the grazing process, and the tracking of egg and juvenile development throughout time lags that may become quite long, necessitated reasonably sophisticated algorithms. Further complexity was required due to a more precise predictor–corrector integration method (see Chap. 7), which separates the estimation of the preferred ingested ration from the metabolic use of the assimilated ration in the flow of the program. In this chapter, the conceptual rationale for the formulations will be presented, followed by a specific account of the complications required to implement it.

The first decision in the specification of this compartment, and one that underlies all other aspects, concerns simply what is to be represented. The major herbivores in Narragansett Bay are copepods, typically dominated by the *Acartia* species *A. tonsa* and *A. clausi*, which may represent 95% of the total population (Martin, 1965; Jeffries and Johnson, 1973). Seasonally, meroplankton such as mollusc larvae, or microzooplankton such as tintinnids, complement copepod grazing. But for the purposes of the model, the zooplankton compartment represents only the small, primarily herbivorous, copepod component of the estuarine system (Fig. 24).

The nutrition of the copepods is complicated by the fact that they apparently consume not only phytoplankton but also other smaller zooplankton (Petipa, 1959, 1966; Hodgkin and Rippingale, 1971) and non-living particulate material (Petipa, 1966; Conover, 1964; Corner, 1972, Gerber and Marshall, 1974). Cannibalism is directly included in the model, with adults consuming eggs and juveniles of the compartment in a density-dependent manner identical to that for phytoplankton grazing. Consumption of particulate organic matter is also provided for, although present lack of data does not allow rigorous use of this option. Some model results suggest that this additional food may play an important role in zooplankton nutrition during critical times of the year, an area touched on by Gerber and Marshall (1974) that needs additional research.

It is assumed that the zooplankton feed unselectively on all available food. While there is evidence sugesting this is the case for estuarine species (Heinle, personal communication, 1974; Gauld, 1951) there is also increasing evidence for

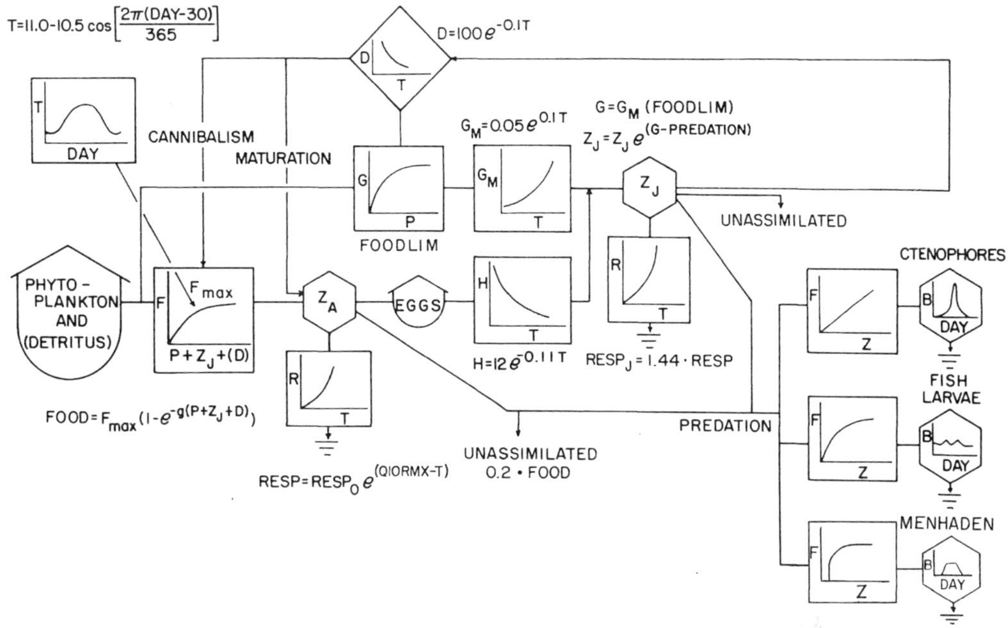

Fig. 24. Flow diagram for the zooplankton compartment showing relationships among the major equations described in text and a graphical representation of their behavior

selectivity based on various criteria, such as particle size (Parsons et al., 1969; Frost, 1972), species characteristics (Gauld, 1951) or relative biomass provided by population components (Berman and Richman, 1974; Heinle, personal communication, 1975). Selection of food type based on characteristics such as these probably occurs and may play an important part in shifts of species dominance associated with succession on various time scales. With the homogeneous phytoplankton community of this model, however, selection is ignored with the assumption that the zooplankters are effectively adjusting their selective bias to the phytoplankton that are dominant at any time. This is consistent with recent experiments of Heinle and Richman that have suggested mechanisms in operation that permit the herbivores most successfully to exploit the dominant primary producers.

The final general consideration concerns the scale of resolution depicted in the zooplankton compartment. Earlier models have treated the zooplankton as a homogeneous pool, growing at a generalized rate representing a composite of juvenile development and adult reproduction (DiToro et al., 1971; Walsh and Dugdale, 1971). A major disadvantage of this is that it ignores time lags resulting from reproductive delays which may be critical in predator–prey cycles, as has been demonstrated by numerous theoretical analyses of the Lotka-Volterra type. Such lumping also precludes any analyses of population structure.

For these reasons, it was desirable from the outset to represent some separation of adult and juvenile functions in the model. The simplest version might be a two-component system, with adults reproducing, and juveniles growing to maturity. Or, perhaps a bit more realistically, one might propose a system with separate pools for adults, eggs, and nauplius and copepodite developmental stages. Technically, however, these intermediate schemes present substantial difficulties. At a given instant, the biomass within a compartment—for example, Stage II copepodites—represents individuals in a continuum of developmental progress. Some may molt to Stage III today, others have a number of days to go. In order to account properly for the progressive advancement from newly released egg through sexually mature adult, each day's stock must be maintained separately within the computer. For cold-water hatching times of 10 days and development times of 90 days, a hundred separate compartments must be tracked, each growing, being grazed by carnivores, etc., for all formulated processes. While this complexity may at first appear to be computer artifact, it should be emphasized that development is more precisely viewed as a continuum despite the discrete jumps which crustaceans undergo. For purposes of analysis, the daily compartments may be grouped in proportion to the relative duration of the developmental stages, providing detailed population structure more directly comparable to natural observations. In most cases, only adults and juveniles have been routinely separated, but more resolution may be readily implemented (see Fig. 53).

In summary, the zooplankton compartment of the model represents omnivorous estuarine copepods. Development is resolved into (1) eggs, which hatch after a temperature-dependent incubation, (2) juveniles, which grow during a temperature-dependent development time, and (3) adults, which reproduce if assimilation is sufficient to meet respiration. Adults are subject to predation by carnivorous zooplankton, and in turn, join these carnivores in exerting grazing pressure on incubating eggs and a proportion of developing juveniles. Metabolic considerations for adults and juveniles include temperature-dependent respiration and excretion of nutrients in varying ratios, temperature and food-density dependent ingestion, and resultant net growth or reproduction.

5.2 Ingestion

Evaluation of the daily ration ingested by the zooplankton is a crucial step in the model, since it is the crux of our understanding of their growth kinetics and their impact on primary producers. Detailed analysis and discussion of this topic has been abundant in the literature for many years, and only a brief synopsis will be related here. As is frequently the case, early disagreement is being resolved by the understanding that no simple generalizations are sufficient to explain all circumstances in nature. This is especially true with respect to the controversy concerning the feeding mode of copepods (see Conover, 1964, 1968; Corner, 1972). Earlier conceptions viewed copepods as exclusively passive filterers, ingesting all capturable material in a specified "volume swept clear". The consequences of this independence of feeding on food concentration seemed consistent with some feeding experiments and with reduced assimilation efficiencies to be expected with proposed superfluous feeding (Beklemishev, 1962). Observations of the filtering rate ranged from a few milliliters per copepod per day up to hundreds, and Cushing (1964) even indirectly estimated a rate of 7 liter daily for *Calanus* V from growth rates during a bloom in the North Sea, a number at odds with even the highest direct measurements (Conover, 1964).

It still appears possible that superfluous feeding may occur in extremely dense algal stocks resulting in destruction of algae that may not even be ingested, perhaps explaining early feeding observations on dense cultures. But it is generally accepted that in the range of phytoplankton concentrations occurring naturally, including fairly dense blooms, feeding is food-density dependent (Conover, 1964; 1966; Frost, 1972). Some high filtering estimates may even be simply due to extrapolation for short-term observations with previously starved animals to full-day rates. Most direct observations on filtering suggest rates of a few ml/day per copepod, with variations of course due to experimental methods and organism size, etc. Interestingly, population metabolic estimates indicate that very low filtration rates often may be sufficient in nature.

It has become increasingly obvious that simple passive filtering is not the sole, or perhaps even a common, feeding mode of primarily herbivorous copepods. Certainly intermittent feeding occurs (Gauld, 1951) and food size and type variations have often been recognized (see Marshall and Orr, 1955a), along with the density dependence of feeding. Alternative feeding modes, such as attack and capture of moving prey (Beklemishev, in Conover, 1964) and encounter feeding, have been cited. The latter, proposed and strongly advocated by Cushing (1959, 1964, 1970), suggests that copepods sense a food particle by contact with antennal spines before moving to consume it. This theory led to an elaborate mathematical formulation (Harris, 1968; Cushing, 1970). Encounter feeding may well occur (see Petipa, 1959; Conover, 1968), despite anatomical objections like the orientation of antennal spines and the ability of some *C. hyperboreus* lacking antenna to successfully feed (Conover, 1964). As has been concluded before (Conover, 1964; Corner, 1972), filtering, active hunting, and encounter feeding probably all occur under different natural conditions of food size and abundance. For the purposes of this model, the exact mode is not important. It is the food-density dependence of ration, and therefore of filtering rate which is consistent with all three modes, that is of direct concern.

The principle concept is intuitive and widely applicable. Utilization of a resource, whether it be energy or nutrients in the case of plants, or consumed food in the case of animals, does not increase indefinitely in proportion to increasing supply. At some point satiation or saturation occurs, resulting in a constant uptake at higher concentrations. In some cases, the plateau drops off at very high levels, as in the photosynthesis–light response, and in some copepod feeding results (Mullin, 1963; Smayda, 1973a). This uncoupling of resource supply and utilization rate may be gradual or distinct. In copepod feeding, detailed observations support both contentions. Frost's (1972) data suggest a fairly distinct transition, implying that feeding continues at a maximum rate at all food concentrations less than that necessary to provide a complete daily ration. Other observations seem to confirm a smooth approach to the maximum ration plateau. In practice, the distinction is small, and the precision of most feeding data is not sufficient to make a definitive decision (Mullin et al., 1975). Figure 25 shows the relationship between ingested ration and filtering rate for the two views. The gradual transition was chosen for this model because a mathematical formulation exists that has been widely used in the

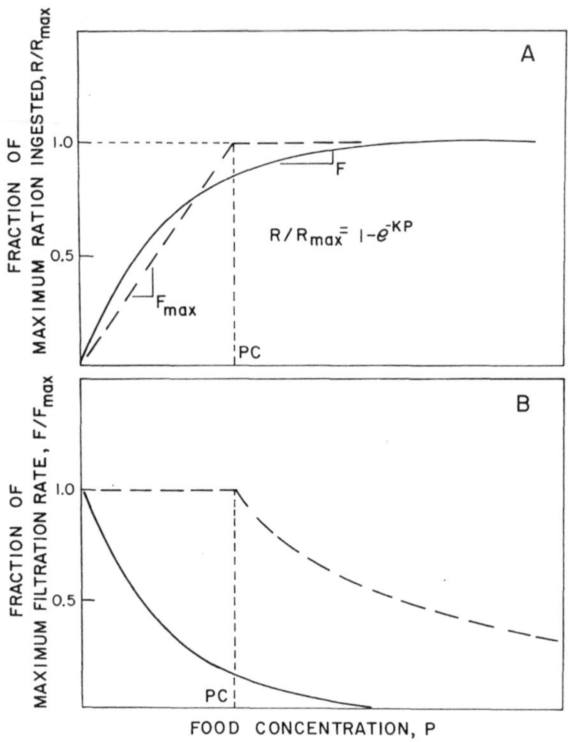

Fig. 25 A and B. Comparison of two alternative formulations representing the saturation of zooplankton feeding at high food concentrations. (A) Two intersecting lines *(heavy dashes)* would be expected to describe ingestion for a passive filter feeding mode, while a hunt-capture-ingest "cycling receptor" mode predicts a hyperbolic relationship such as that of Ivlev *(solid curve)*. (B) Decrease in filtering rate at high food levels associated with the two schemes. In the linear case, filtration continues at a maximum rate up to a critical food concentration (*PC*)

interpretation of feeding observations. However, before completing the discussion of the exact formulation, a brief theoretical analysis of some implications and relationships of the alternative interpretations is appropriate.

5.2.1 Ration, Filtering Rate, and Experimental Observations

The interpretation of Frost (1972) divides feeding into two regions: (1) where food is insufficient to provide the preferred ration and filtering occurs at some constant rate presumably approaching the organism's maximum, and (2) where food densities saturate the feeding rate and the ration may be achieved at lower filtering rates or at maximum rates for shorter time intervals (Fig. 25). Although there is undoubtedly variability in the switching point, resulting in some smoothing of the transition, the logic seems ecologically reasonable especially assuming a feeding mode of roughly passive filtering. The achieved ration (R), and the apparent filtering rate (F) are tightly related, either being determinable from the other plotted against food density. However, the slope of the ration line defines F only below the critical food concentration, going to zero above PC (Fig. 25A). Thus:

$$\text{Slope} = \frac{dR\,(\text{mgC cope}^{-1}\,\text{day}^{-1})}{dP\,(\text{mgC l}^{-1})} = F\,(\text{l cope}^{-1}\,\text{day}^{-1}).$$

A similar analysis is possible for the curvilinear case. The most commonly used mathematical expression is that suggested by Ivlev (1945) to express the hyperbolic pattern of juvenile fish ingestion as a function of available food:

$$R = R_{max}(1 - e^{-k \cdot P}). \tag{18}$$

In this equation, R_{max} determines the units, e.g. mgC cope^{-1} day^{-1}, since the exponent k (l/mgC) has the units of inverse food concentration P (mgC/l). The equation is a case of a general form (Dr. Ian Fletcher, Univ. Washington, personal communication):

$$y = \frac{c}{b}(e^{-bx} - 1).$$

The differential equation for this is:

$$y' + by + c = 0 \tag{19}$$

and at $y=0$

$$y' = -c.$$

For the Ivlev equation, we must substitute:

$$R_{max} = \frac{-c}{b}$$

$$k = b$$

$$R_{max} \cdot k = -c. \tag{20}$$

And finally, we may conclude that the slope at the origin $P=0$ represents an estimate of the maximum filtering rate. Since $y' = -c$:

$$\frac{dR}{dP} = R_{max}\left[\frac{\text{mgC}}{\text{cope}\cdot\text{day}}\right] \cdot k\left[\frac{1}{\text{mgC}}\right] = \hat{F}\left[\frac{1}{\text{cope}\cdot\text{day}}\right]. \tag{21}$$

In this approach, the slope of the curve continues to give an estimate of F for all food concentrations:

$$y' = -by - c$$

$$\hat{F} = -R \cdot k + R_{max} \cdot k \qquad \text{[see (19)]}$$

$$\hat{F} = k(R_{max} - R).$$

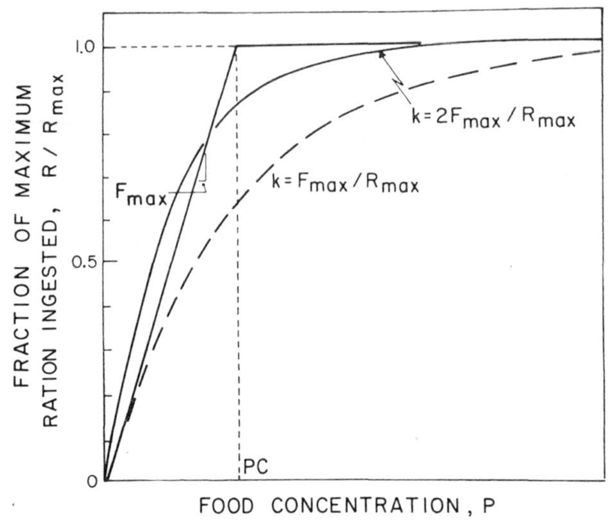

Fig. 26. Empirically derived relationship between a maximum filtering rate and the appropriate exponent (k) in the Ivlev Eq. (18). Specifying $k = 2F/R_{max}$ provides a more satisfactory fit to the initial maximum filtering rate below the critical food concentration (PC) where the maximum ration is predicted

A weakness here is that $\hat{F}$ declines for any increase in food density, even for P near zero. Also, the relation is asymptotic and high values of k are necessary for R to approach a high fraction of R_{max} at realistic food concentrations. Thus a choice of the Ivlev k based on F divided by R_{max} [from Eq. (21)] does not achieve R_{max} very rapidly (Fig. 26). Empirically, doubling F for an estimate of k results in a more satisfactory pattern. By comparison, at the point where Frost's linear interpretation reaches the maximum, a ration of about 90% R_{max} is predicted using $k = 2F/R_{max}$ in contrast to 60% for $k = F/R_{max}$. In the interpretation of feeding data reported in the Ivlev format, then, an approximation of half the suggested initial slope may represent an estimate of the maximum filtration rate:

$$F = 0.5\hat{F} = 0.5 R_{max} \cdot k. \tag{22}$$

Two cautions in the application of this theoretical development should be emphasized. First, statistical methods of fitting equations like (18) to linearized data may weight the points with greater numerical value more heavily, thus biasing against the low food concentrations which are most interesting in evaluating filtering rates. Second, as mentioned in the discussion of the photosynthesis–light formulation (Chap. 4), analysis of such responses must be done with unnormalized data. When the unitless R/R_{max} is plotted with food density, it is impossible to distinguish apparent changes in k from differences resulting from unequal R_{max} values.

Finally, it should be emphasized that both the hyperbolic and linear formulations are consistent with zooplankton feeding modes known to occur. Frost (1972) and Mullin et al. (1975) have pointed out that a mode of more or less passive filtering would probably demonstrate the linear pattern. Other modes, such as encounter feeding or active hunting (see Conover, 1964; Cushing, 1970) are examples of a theoretical "cyclic-receptor" (Odum, 1972). Both Holling (1966, 1969) and Odum have demonstrated that a hyperbolic pattern should result from this type of mechanism in a manner directly analogous to the Michaelis-Menten model of enzyme kinetics. It seems reasonable that in nature or experimental situations, ingestion by copepods may be the result of one or more of these modes, additionally complicated by temporally intermittent rates. A precise interpretation may not be possible, and certainly the practical consequences of either mode are similar in terms of experimental comparison or modeling.

5.2.2 The Maximum Ration

The approach chosen in the model was to specify the maximum ration of the zooplankton as a function of temperature. This is then reduced by food-density dependent limitation and any competition resulting from dense biomass. While ration estimates may be achieved through a direct empirical filtration–temperature relation (DiToro et al., 1971), this is often difficult or tenuous, perhaps due largely to the considerations of the previous theoretical discussion. Some extensive filtration observations even omit entirely any mention of food concentration (Anraku, 1964). Estimates of ration reported as such, while no less variable, are more directly applicable to the goals of the model, and thus were used in the ingestion formulation. A brief comparison demonstrates the consistency of the two sets of observations.

For reasonable ranges of copepod abundance in Narragansett Bay of 20–70 animals/l and phytoplankton standing stocks between 0.001–0.01 mgChl/l, approximate rate estimates necessary to provide a daily ration of 50% body carbon can be calculated. Assuming 0.006 mg dw/copepod, 35% Carbon, and phytoplankton C : Chl = 50:

			Filtration	
Zoopl No/l	Phyto mgC/l	Total ml/day	per individual ml/(cope·day)	Wt specific ml/(mg dw · day)
20	0.05	420	21	3500
	0.50	42	2	350
70	0.05	1470	21	3500
	0.50	147	2	350

For simplicity, the exponential nature of filtering has been ignored, and clearly it is impossible for all 70 copepods to attain their entire ration at the lower food level. Nevertheless, the comparison of the filtrations per copepod and per milligram compare favorably with many literature estimates (Gauld, 1951; Marshall and Orr, 1955a; Conover, 1956, 1968; Anraku, 1964; Martin, 1966).

The daily ration presumably reflects the basic metabolic needs of the organism. Therefore, a good working hypothesis is that the temperature response will follow an exponential relation according to the van't Hoff rule. While seasonal acclimation and succession complicate the pattern, the general form of the relationship is well established:

$$R_{max} = R_{max_0} e^{(Q10RMX \cdot Temp)}, \tag{23}$$

where: R_{max_0} = the maximum ration at 0° C, $\frac{mgC_{ingested}}{mgC_{zoo} \cdot day}$,

Q10RMX = temperature rate constant ($°C^{-1}$).

Note that the ingestion and biomass units of R_{max} are the same, resulting in a dimension for R_{max} of "per time", an instantaneous ingestion rate. Q10RMX is from the logarithm of the physiological Q_{10}; e.g., for $Q_{10}=2.0$, Q10RMX=(ln 2.0)/10=0.069° C^{-1}.

Figure 27 for the realized ration of adult *Acartia clausi* (Petipa, 1966) demonstrates this relationship. Measurements were for fully fed animals and thus represent something near R_{max}. The exponent of 0.12 represents a high $Q_{10}=3.3$. Observations on *A. clausi* from Narragansett Bay (Smayda, 1973a) showed a $Q_{10}=1.6$ between 4–10° C in the laboratory.

The great range in measurements of R_{max} probably represents at least in part real variation, even beyond differences due to experimental technique, food type, organism size and location, and so on (Fig. 28). There are even recent suggestions that the upper limit can vary with acclimation of copepods to ambient food levels. The two parameters controlling ingestion, R_{max} and Q10RMX, are therefore left

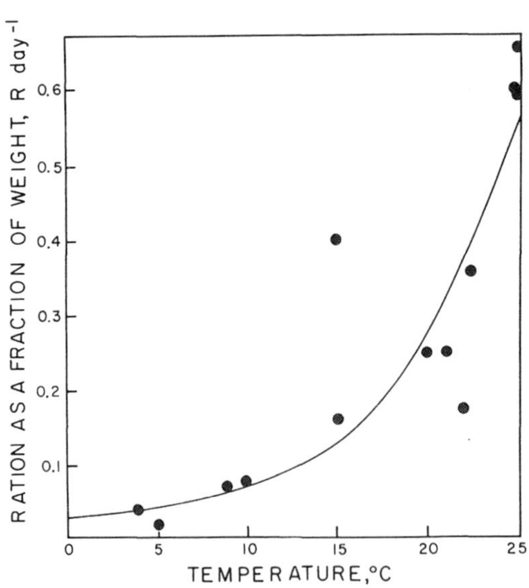

Fig. 27. Daily ration of adult copepods as a function of temperature. Data for *A. clausi* from Petipa (1966) with least squares fit $R=0.024e^{0.12 \cdot TEMP}$

unassigned in the program, and one is free to specify various combinations. The sensitivity to these critical coefficients will be discussed in Chapter 11.

5.2.3 Reducing the Maximum Ration

In nature a myriad of factors may result in zooplankters actually ingesting less than the potential maximum ration. Competition, food density, food type, size and nutritional value all may affect the realized consumption. In the model only two are considered: supply of available food and competition due to high densities of copepods.

The hyperbolic nature of ingestion with increasing food concentration has already been mentioned. In the model, the mathematical formulation of Ivlev (1945) was selected to represent this trend, as given earlier [Eq. (18)] and restated here in the model notation:

$$\mathrm{XRTN} = 1.0 - e^{-\mathrm{IVLEVK} \cdot \mathrm{PCMG}}, \tag{24}$$

where: IVLEVK = empirical constant controlling the degree of density-dependent limitation (l/mgC),

PCMG = food concentration, particulate carbon (mg C/l),

XRTN = fraction of maximum ration.

In this normalized form, XRTN is a unitless fraction, representing the ratio of the realized ration to the maximum. The exponent IVLEVK determines the steepness of the curve approaching 100% of R_{max}. With units the inverse of food concentration, IVLEVK represents the reciprocal of that density resulting in 63.2% of the maximum ($1.0 - e^{-1} = 0.632$). Higher values of IVLEVK cause a larger realized ration to be achieved at lower densities. Figure 28 presents curves fitted to feeding data of various workers by nonlinear least squares regression with Eq. (18). The original data were reported in various units, and conversions to standard carbon values were necessary (Table 8). Nevertheless, substantial variations in the exponent exist, and this parameter may be varied appropriately in the computer program.

Two limitations of the ingestion formulation are pertinent. Nocturnal grazing has been postulated for many zooplankton species, usually related to vertical migration patterns, McAllister (1970) has presented a strong argument for the importance of this in altering both the magnitude of the ingested ration, and the resulting impact on the algal population. In a series of controlled grazing experiments coupled with a modeling analysis, it was demonstrated that periodic grazing increased ingestion by the copopods, presumably due to overfeeding after a period of starvation, yet minimized the depression of algal growth due to the timing of cell division. Despite this, in the present model such periodicity was not included. The 24 h time-step requires that appropriate average rates be used. More importantly, it is likely that copepods in shallow estuarine environments do not exhibit strong vertical migration. Petipa (1967, in Mullin, 1969) observed Black Sea

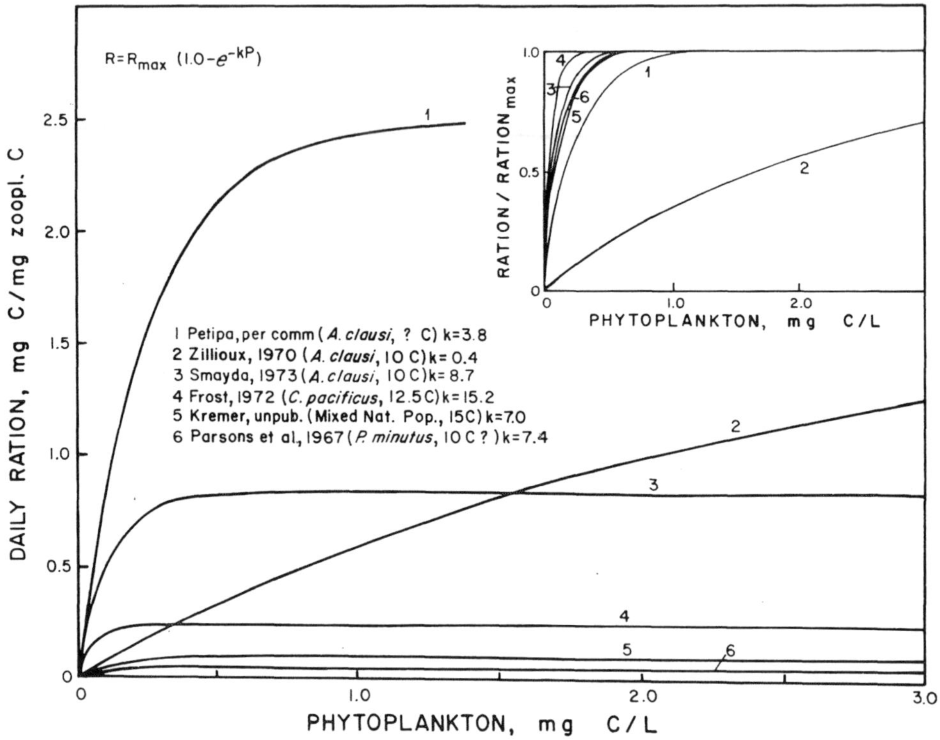

Fig. 28. Comparison of experimental results relating food concentration to daily ration of zooplankton. Original data have been converted to standard units as accurately as possible (Table 8) and fit to the feeding equation of Ivlev [Eq. (18)]. Insert represents the ration normalized with respect to the maximum, allowing direct comparison of the exponent, k. (From Kremer and Nixon, 1975)

A. clausi to swim continuously and feed in the upper 20 m with no diel periodicity, although earlier work had suggested temporal partitioning might separate adult and juvenile grazing somewhat (Petipa, 1958).

The second consideration in the use of the Ivlev relation is the question of a threshold for zooplankton feeding. Some workers (Petipa, personal communication; Parsons et al., 1967, 1969; McAllister, 1970; Frost, 1975; Mullin et al., 1975) have clearly measured food concentrations below which feeding in the laboratory appears to be negligible. Other observations depict a strong trend for feeding linearly related to food at all low concentrations (Smayda, 1973a; Frost, 1972). Even though the empirical evidence is conflicting, the inclusion of a threshold is mathematically straightforward (Parsons et al., 1967):

$$\mathrm{XRTN} = 1.0 - \mathrm{e}^{-k(\mathrm{PCMG} - P_0)}, \tag{25}$$

where: P_0 = a lower feeding threshold (mgC/l).

When food concentration is below P_0, the exponent becomes positive driving XRTN negative, necessitating a programming correction to zero predicted ration.

Table 8. Some experimental measurements of R_{max} and Ivlev-k [see Eq. (18)]. Reported data have been converted to standard units, R_{max}(mgC mgC^{-1} day^{-1}), k (l mgC^{-1})

Reference	Conditions, comments	Reported R_{max}, and P units	Fitted R_{max}	Fitted k
1. Petipa (personal communication)	*A. clausi*, ?C	R_{max} = 0.044 Cal copepod^{-1} day^{-1} P in Cal m^{-3}; P_0 = 126 cells/ml	2.50	3.8
2. Zillioux (1970)	*A. clausi*, 10C, fed *Rhodomonas baltica*; ^{14}C-labeling; 1 h expts; 100 ml vol?	R_{max} = 1600 cells copepod^{-1} h^{-1}; P in cells/l	1.8	0.4
3. Smayda (1973b)	*A. clausi*, 10C, fed *Skeletonema costatum*; 100 copepods l^{-1}	R_{max} = 5.8 × 10^4 cells copepod^{-1} h^{-1}; P in cells/l	0.83	8.7
4. Frost (1972)	*Calanus pacificus*, 12.5C; continuous light; Fed various diatoms in open and closed systems	R_{max} = 1.1 μg C copepod^{-1} h^{-1}; P in μg C/l.	0.25	15.7
5. Parsons et al. (1967)	*Pseudocalanus minutus*, ?C, fed natural phyto.; 24 h expts, in dark; 850 ml vol	R_{max} = 1.0 × 10^6 μ^3 copepod^{-1} day^{-1}; P in μ^3/ml; P_0 = 5.8 μg C/l	0.05	7.4
6. Suschenya (1970)	*A. clausi*, 20C, fed various algae	R_{max} = 0.0011 mg wwt copepod^{-1} day^{-1}; P in mg wwt/l	0.013	25.0
7. McAllister (1970)	*Calanus pacificus*, LD cycle, continuous and periodic grazing	R_{max} = 0.018 μgC μgC^{-1} h^{-1} (12 h); R_{max} = 0.008 (24 h); P in μg C/l; P_0 = 15 μg C/l.	0.22 0.19	10.0 14.0
8. Kremer (unpublished)	Mixed natl. copepods, 15 C, fed S. *costatum*	R_{max} = 6600 cells copepod^{-1} day^{-1} P in cells/ml	0.10	7.0

Walsh (1975) has justified the inclusion of a grazing threshold on completely different grounds. He considers the threshold "essentially a parameterization of the inability of the herbivore submodel to resolve horizontal scales of prey patchiness" smaller than the grid size of the model. Thus zooplankton, or herbivorous anchovies in Walsh's case, are simply assumed not to be found in regions of sparce food. In the very large elements of the Narragansett Bay model, to effectively exclude zooplankton from entire elements would be unsatisfactory. Moreover, while the interpretation may have some validity for nekton, it goes against the observations of phytoplankton–zooplankton exclusion, which have puzzled plankton ecologists for many years. Rather, by postulating feeding with no arbitrary exclusion, we may observe whether such inverse patterns may develop in the model system.

Mathematical models do not necessarily honor the "positivity constraint" existing in the real world, and the threshold formulation may be useful in avoiding overgrazing of phytoplankton which may result artifically from finite difference approximations. Grazing thresholds were not included in our model because it is not clear that in nature such a distinct cutoff generally exists. In addition, the numerical treatment of grazing developed here results in an effective threshold which may more realistically reflect the depression of feeding at low phytoplankton concentrations. This relates directly to the final point in the discussion of nonoptimum feeding in the model, competition among abundant zooplankton.

In a section on mathematical considerations (Chap. 7), the method of handling instantaneous rates in a large time-step finite difference scheme will be presented in detail. Briefly, if it is assumed that the absolute rates are constant during a time interval, the explicit integral may be specified with no numerical error. Thus, for a grazing rate, F, acting on an initial phytoplankton concentration, P_0, explicit integration is according to the familiar equation.

$$P_t = P_0 \cdot \mathrm{e}^{-F \cdot \Delta t}$$

in comparison to a differencing approximation given by:

$$\Delta P = P \cdot F \cdot \Delta t$$

$$P_t = P_0 + \Delta P .$$

In addition to the fact that any number of independent filtering rates may be added together simply in the single value F, a desirable consequence of this scheme is that exponential grazing evaluations will never drive the algae to zero. The *intensive* (Callen, 1963; Slobodkin, 1975) property of concentration "sensed" by individual zooplankters is successfully evaluated in its *extensive* context: is the total phytoplankton biomass sufficient to meet the combined demand of the grazing community?

For many interpretations of zooplankton feeding mode, the instantaneous filtering rate is theoretically appropriate. Indeed, zooplankton grazing data are often reported as rates based on the appropriate logarithmic equation [see Eq. (41)]. In the model, the potential ration is the quantity that is initially calculated.

Based on this value, the preferred ration for the total zooplankton standing stock is projected, and this estimate is used to derive an instaneous filtering rate, liters cleared per liter per day.

Let

XRTNMX = maximum ration of zooplankton as a fraction of body weight [Eq. (23)] ($\mathrm{mgC_{ingested}\,mgC_{zoo}^{-1}\,day^{-1}}$),

XRTN = food limitation term [Eq. (24)],

Z = zooplankton concentration (mgC/l),

P = available food concentration (mgC/l).

Then RTN = XRTNMX · XRTN · Z, the *estimated* ration of the total zooplankton stock ($\mathrm{mgC\,l^{-1}\,day^{-1}}$),

and $\mathrm{RLF} = \dfrac{\mathrm{RTN}}{P}$, the instantaneous filtering rate ($\mathrm{day^{-1}}$).

RLF, the ration as liters filtered, represents the rate at which each zooplankter must filter in order to achieve its preferred ration if the food concentration does not change. This is an *intensive* property based on the perception of each individual, without regard to other effects such as competition. This rate is then evaluated as in Eq. (49) resulting in some impact on the phytoplankton, with a certain *realized* ration actually becoming available to the zooplankton. The realized ration will rarely equal the estimated ration, but the difference is small as long as the rates are not extreme. As competition for food increases, for example, when zooplankton standing stock rises, RLF may approach or exceed 1.0, and competition will diminish the realized ration with respect to the preferred ration.

5.2.4 Cannibalism

The importance of animal material in the diet of copepods that have traditionally been considered herbivorous is now well established (Petipa, 1959, 1966; Hodgkin and Rippingale, 1971). These findings have been included in the model by allowing the adults of the zooplankton to feed on the eggs and juveniles of the compartment. Feeding on these sources is not functionally distinguished in the model—the density dependence and the lack of selectivity are the same as for phytoplankton and nonliving particulate organic carbon consumption. Because larger stages of developing copepods might reasonably be less subject to adult predation, the model provides for a variable portion of the juvenile compartment to be exempted from cannibalism.

In the model symbolism, SAFE designates the fraction of development during which juveniles are large enough to avoid cannibalism. SAFE = 1.0 eliminates cannibalism entirely. The adult's ration of juveniles, RTNZJ, is calculated by a simultaneous evaluation of projected juvenile growth and the combined carnivorous filtering impact, according to Eq. (49). The consumption of eggs, RTNEGG, may be more simply evaluated by the exponential loss due only to the filtering rate since the eggs are not growing during the incubation lag.

The complete ingestion formulation for adult zooplankton, then, uses the following equations:

$$\begin{aligned}
\mathrm{XRTNMX} &= \mathrm{RMX0} \cdot \mathrm{EXP}\ (\mathrm{Q10RMX} \cdot \mathrm{TEMP}) \\
\mathrm{RTNMX} &= \mathrm{XRTNMX} \cdot \mathrm{ZAMGC} \\
\mathrm{PCMG} &= \mathrm{PTOT} + \mathrm{POM} + \mathrm{EGGTOT} + \mathrm{ZJCANN} \\
\mathrm{XRTN} &= 1.0 - \mathrm{EXP}(-\mathrm{IVLEVK} \cdot \mathrm{PCMG}) \\
\mathrm{RTN} &= \mathrm{XRTN} \cdot \mathrm{RTNMX} \qquad \text{(estimated)} \\
\mathrm{RLF} &= \mathrm{RTN}/\mathrm{PCMG}
\end{aligned}$$

Based on these values, growth and combined grazing rates are evaluated simultaneously, specifying the components of the adult ration so that:

$$\mathrm{RTN} = \mathrm{RTNP} + \mathrm{RTNZJ} + \mathrm{RTNEGG} + \mathrm{RTNPOM} \qquad \text{(realized)}. \tag{26}$$

RTN, the daily ration of the adult zooplankton compartment in mgC (implicitly $\mathrm{mgC\,l^{-1}\,day^{-1}}$), is first estimated based on the preferred ingestion and the density limitation. ZJCANN is the biomass of juveniles subject to adult predation, POM the concentration of edible detrital carbon, and EGGTOT the biomass of zooplankton eggs. RLF is a filtering rate ($\mathrm{day^{-1}}$) which would provide this ration to the entire community if the food concentration stayed constant. RLF may exceed 1.0 when high zooplankton and low food densities result in excessive grazing pressure. RLF is then evaluated simultaneously with other grazing pressures and growth rates for the phytoplankton [see Eq. (49)], resulting in calculation of the ingested plant carbon (RTNP). Other components of the ration—cannibalized eggs (RTNEGG), susceptible juveniles (RTNZJ), and other particulate organic matter (RTNPOM)—are combined to yield the realized ration that is ingested and available for assimilation.

5.3 Assimilation

The fraction of the ration ingested by copepods that is actually available for metabolic work has been widely studied. Historically, the accepted view has changed, associated with a more complete understanding of the complexities of zooplankton feeding mode. The early hypothesis that copepods are constant-rate filter feeders suggested that rations far exceeding the maximum demands of the organisms may be achieved when food is abundant. This view was explicitly stated in the superfluous feeding theory of Beklemishev (1962), and was actively debated in the literature for a number of years (see Conover, 1966).

According to this theory, excessive zooplankton ingestion at food concentrations exceeding some upper threshold would be accompanied by decreasing assimilation efficiencies. Such behavior would seem to be unnecessarily destructive of phytoplankton, though enhanced nutrient recycling as well as rich feces production, providing an energy resource for the ecosystem, might be beneficial consequences (Cushing and Nicholson, 1963; Conover, 1964).

In conjunction with increasing evidence that copepod feeding is usually food-density dependent, better estimates of assimilation efficiency (assimilation/ingestion) have become available. Most of these data do not support the superfluous feeding theory. Rather, no clear relation—inverse or otherwise—between assimilation efficiency and food concentration is usually evident (Conover, 1964, 1966). Perhaps the fairest assertion at this point is that although excessive destruction of algal cells with or without accompanying ingestion may occur under some circumstances (see Cushing, 1964), for the range of phytoplankton densities naturally encountered, assimilation is high and relatively constant (Corner, 1972). Based on this, the model assumes that a constant fraction of the ingested ration is assimilated by the zooplankton. Numerous measurements employing a variety of methods are available to suggest an appropriate range of efficiencies. Some of the earliest direct measurements on *C. finmarchicus* (Marshall and Orr, 1955, 1955a) demonstrated efficiences consistently higher than about 60%, and predominantly above 80%. Later work has continued to support these findings (Conover, 1964, 1966, 1966a, 1968; Petipa, 1966; Haq, 1967; Butler et al., 1969; Mullin, 1969).

Conover (1966) has discussed a number of other factors potentially affecting assimilation which deserve mention. Different phytoplankton species resulted in assimilation efficiency differences that were apparently inversely related to the ash content of the algae. Within a single species culture, age had no demonstrable effect. Temperature had no significant effect in a preliminary series of experiments, and even during a five-day transition from an inactive, resting state to one of active feeding, *C. hyperboreus* maintained an approximately constant assimilation efficiency while more than doubling ingestion. In summary, while assimilation efficiency is certainly variable, with the possible exception of nutritional value (ash content), no clear patterns have emerged suggesting which external factors may be related to the differences. At least within the context of the generalized phytoplankton compartment of the model, the simple assumption of a constant assimilation efficiency is justified. For most simulations, a value of 80% was used.

Since carbon is the basic unit of biomass of the model, the designated efficiency is assumed to refer to carbon assimilation. However, the assimilation of nitrogen and phosphorus must also be considered. Many of the observations referred to above in support of a high-assimilation efficiency were based on measurements of nitrogen or labeled phosphorus, and the generally high efficiency is widely consistent (see also Butler et al., 1969).

In nature, and in the model, zooplankton may have C:N and C:P ratios different from their food. Therefore it was necessary to formulate a scheme allowing the animals to maintain their nutrient ratio regardless of the ratio of the food. While most of this balance is maintained through variations in nitrogen and phosphorus excretion, if the food is extremely nutrient deficient, a provision for retention of nutrients from the normally unassimilated part of the ration is included. The details of this formulation are presented in the section on nutrient excretion in Chapter 6.

The unassimilated nutrients (UAN, UAP, UASI) are calculated in the model from the combined zooplankton ration, which may consist of phytoplankton, detritus, and cannibalized juveniles and eggs.

$$UAN = XUNASM \cdot [N{:}CP \cdot TOTRP + N{:}CPC \cdot TOTRPM + N{:}CZ \cdot (RTNZJ + RTNEGG)]$$

$$UAP = XUNASM \cdot [P{:}CP \cdot TOTRP + P{:}CPC \cdot TOTRPM + P{:}CZ \cdot (RTNZJ + RTNEGG)]$$

$$UASI = SI{:}CP \cdot TOTRP + SI{:}CPC \cdot TOTRPM, \tag{27}$$

where: TOTRP = adult + juvenile ration of phytoplankton,
TOTRPM = adult + juvenile ration of particulate matter,
RTNZJ + RTNEGG = adult ration due to cannibalism,
XUNASM = the fraction of the carbon ration not assimilated = 1. − ASSIM,
N:C, P:C, SI:C = nutrient–carbon ratios for phytoplankton (P), zooplankton (Z) and particulate carbon (PC).

Silica content of zooplankton is assumed to be zero, so all of the ingested silica is released as UASI. Unless some of these losses are required to maintain zooplankton tissue composition, the unassimilated nutrients are lost to the benthos as sinking fecal pellets.

In nature, a portion of the nutrients in zooplankton fecal pellets is regenerated into the water column during the time it takes for them to sink to the benthos, although this process may be inhibited by tight organic binding of the nutrients (Lewin, 1961). In addition, pellets are probably available for reingestion by the zooplankton. Coprophagy may readily be added to the present model by connecting the unassimilated pools of nutrient, with an additional carbon component, to the particulate carbon compartment. Except for this feedback, the necessary model coding already exists. However, neither coprophagy nor fecal nutrient regeneration in the water column have been included in any of the simulations that will be discussed in this book.

5.4 Respiration

In energetic terms, respiration is the loss to heat (or entropy) required by the Second Law of Thermodynamics to accompany any transformation of energy. This metabolic tax must be paid before any useful work can be done, such as physical movements, growth, or reproduction. As a basic metabolic function, respiration may be expected to be a function of temperature similar in form to that of the maximum ration [Eq. (23)]

$$XRSP = XRSP_0 \, e^{(Q10RSP \cdot Temp)}, \tag{28}$$

where: $XRSP_0$ = respiration rate at 0° C (mgC mgC^{-1} day^{-1}),
Q10RSP = temperature rate constant, $\ln Q_{10} \div 10$ (° C^{-1}).

Strictly according to thermodynamic and biochemical considerations, the Q_{10} of this relationship migh be expected to be about 2.0 (Q10RSP = 0.0693° C^{-1}). But actual observations reveal that the problem is much more complicated in nature. Organism size, physiological state, physical activity, and seasonal acclimation are among the factors that make a simple specification of respiration solely as a function of temperature imprecise (see Marshall et al., 1935; Raymont and Gauld, 1951; Gauld and Raymont, 1953; Conover, 1959, 1960; Conover and Corner, 1968; Ikeda, 1970; Mullin and Brooks, 1970).

For example, short-term changes in respiratory rate are known to occur in relation to both nutritional and activity state. Thus, starving or fasting animals, or those that are intermittently inactive or in diurnal periods of rest, may be expected to have depressed respiration in comparison to normally active animals (Conover, 1968; Ikeda, 1970; Mayzaud, 1973). These variations, of course, are readily explicable by basic physiological considerations. In this model, however, rates are averaged over daily time intervals, and these short-term variations are on too fine a scale to be represented.

Larger organisms obviously respire more than smaller ones, but the weight-specific respiration rate (e.g., mgC respired/mgC · day) is well known to be an inverse function of size (Zeuthen, 1953; Conover, 1968; Corner, 1972). The allometric relationship for absolute respiration is of the form:

$$\text{Resp} = a \cdot w^b \qquad \frac{\text{respiration}}{\text{organism} \cdot \text{time}},$$

where a and b are constants. In the weight-specific form:

$$\text{Resp}' = \frac{a \cdot w^b}{w} = a \cdot w^{(b-1)} \qquad \frac{\text{respiration}}{\text{unit wt} \cdot \text{time}}. \tag{29}$$

As expected from the theoretical analysis of surface-to-volume considerations, b is frequently between 0.67 and 1.0. The homogeneous carbon pools representing the biomass in the model zooplankton compartments are not associated with numbers of individuals, thus precluding any specification of size, and size-variable respiration. The above relation, however, may be used to estimate the average increased respiration of juveniles with respect to adults.

Petipa (1966) presents the weight of *A. clausi* throughout a 30-day development. An average juvenile weight during development may be estimated using her data on the size and duration of the development stages.

Stage	Weight · Duration		
Nauplii	0.01 μg	11 days =	0.11
Copep. I	0.21 μg	3 days =	0.63
Copep. II	0.42 μg	3 days =	1.26
Copep. III	0.73 μg	3 days =	2.19
Copep. IV	1.81 μg	3 days =	5.43
Copep. V	2.56 μg	7 days =	17.92
Total, average		30 days	27.54 ÷ 30 = 0.92 μg

Thus, the average weight during development was 0.9 μg vs. an adult weight of 5.0 μg. For the allometric relationship, $R = aw^b$, Conover (1968) reviews numerous observations which suggest $b \simeq 0.8$ for estuarine copepods. Assuming the proportionality constant a is similar for adults and juveniles, the ratio of the weight-specific respiratory rates [Eq. (29)] can be determined:

$$\mathrm{RJ}:\mathrm{RA} = \frac{R_j}{R_a} = \frac{a(0.9)^{0.8}/0.9}{a(5.0)^{0.8}/5.0} = 1.4\,. \tag{30}$$

Thus, the effect of size on respiration of zooplankton may be represented simply by specifying the juvenile rate as 1.4 times the temperature-dependent adult rate [Eq. (28)]. Although the allometric relation often is used with oxygen consumption, the units of the constant a may be carbon respired per unit carbon of biomass. As long as the same respiratory quotient (RQ) is assumed for adults and juveniles, the ration calculation is valid for the carbon base of biomass used in the model. While this may not be an overly strong assumption considering potentially different food resources and metabolic demands of juveniles and adults, it is probably satisfactory for the present purposes of the model. The above calculation is reasonably sensitive to the choices of initial and adult weight, and considering the real variability as well as the uncertainty in any given estimate, the exact value of 1.4 should not be accepted too seriously. Calculations from respiration and size measurements for two large species of marine copepods under various conditions give a ratio of 1.5 between copepodite Stage IV and adult (from data in Table 4, Mullin and Brooks, 1970; mean of 9 values with one high point discarded). While this would suggest an even higher value for the total development period, the approximate agreement supports the general formulation.

The respiration formulation of the model [Eq. (28)] implicitly considers both the thermodynamic effect of temperature and the seasonal acclimation of the community. Again, because of the generality of the homogeneous carbon compartments, detailed species considerations are not possible, or necessary. Only the resultant pattern of thermal respiratory response is represented, without regard to whether it is due to shifts in species dominance or temporal acclimation, or both, as is usually the case. But these considerations are important in the choice of a numerical value.

According to the conventional approximation, a Q_{10} of about 2 should represent the unacclimated thermodynamic increase in a biochemical reaction rate. Any physiological acclimation or community succession would make the organisms better suited to their environment, presumably decreasing the temperature effect and therefore the Q_{10}. On the other hand, basal metabolic rate is neither the only process reflected in the respiratory rate, nor the only one affected by temperature. As temperatures increase, metabolic or behavioral activities which were insignificant at lower temperatures may begin, contributing to increased oxygen consumption. In other words, the effective Q_{10} over the wide temperature range for these activities might be considerably greater than 2. Thus, active swimming, feeding and digestion, and reproduction might account for higher Q_{10}'s in nature than projected from winter rates and a simple $Q_{10} = 2$, even allowing for some beneficial acclimation (see Conover, 1959). Stress conditions would similarly invalidate any such Q_{10} rule, due to increased metabolic work performed simply as a passive response to the stress, or in an attempt to regulate or protect against it. In view of these opposing arguments, the wide disagreement in Q_{10} reported in the literature is not surprising.

Numerous workers have measured unacclimated Q_{10}'s for respiration of marine and estuarine copepods between 2 and 3 (see Mullin and Brooks, 1970). Some unusually high results are probably due to the stress response mentioned above, although the average values of 3.1 for *C. helgolandicus* and 8.3 for *Rhincalanus nasutus*, measured between 10 and 15° C for fully acclimated lab-raised animals, seem extreme (Mullin and Brooks, 1970). In contrast, seasonal observations of the respiration of natural copepod populations at ambient temperatures seem to suggest that acclimation is effective in achieving ecological Q_{10}'s less than 2 (Conover, 1959; Anraku, 1964; Conover and Corner, 1968). However, this conclusion is universally qualified by the complications of seasonal patterns of activity and life cycles.

Future modification of the respiration formulation might well include a basal rate, temperature-dependent as in Eq. (28) with a relatively shallow Q_{10}, and an additional activity-related component. This rate would be coupled to feeding and ingestion, and perhaps to reproduction as computed in the model. Such a formulation also would permit a more-detailed representation of nitrogen and phosphorus excretion, which is closely related to variations in respiratory rate. For the present model, the more simplified approach is retained as a generalized formulation flexible enough to represent a variety of responses.

5.5 Excretion

In the previous section, respiration was discussed as an indicator of the metabolic work done by the zooplankton. The release of nutrients is as directly related to respiration as the oxygen uptake and carbon dioxide production that are usually used to measure metabolism. For this reason, nutrient excretion in the model is closely tied to the respiration formulation just presented.

The connection between nutrient excretion and carbon respiration depends on the elemental composition of the substrate being utilized in catabolism. Since the nutrient content of the original food and the various storage reserves may differ dramatically, the release of nutrients is quite variable over short time intervals. In addition, the process is complicated by the many factors that may result in variations in respiration, which is the driving rate. At present the basis of the excretion formulation is directed at maintaining appropriate zooplankton tissue composition despite differences in the content of the ingested food, and the resultant variations in nitrogen and phosphorus excretion.

Nitrogen excretion by zooplankton and its potential role in ecosystems has been the subject of a great deal of recent research. Phosphorus excretion, while also widely studied, has not received quite the same attention. For example, critical data on O : P ratios which relate to the substrate being metabolized were not available for the 1971 review by Corner and Davis. The mechanisms involved are likely to be closely related, however, and with the exception of the forms of released compounds, conclusions on excretion in general are probably possible, especially in the broad context of the model.

Experimental observations have demonstrated that nitrogen excretion shows systematic variations with organism size, temperature, and what may be assumed to be metabolic substrate. Taxonomic differences, of course, also are evident, but zooplankton, including copepods, may be satisfactorily classified together as excreting unusually high levels of nitrogen in comparison to other animals, even other crustaceans (Mayzaud, 1973). Prominent seasonal variations haven often been noted, but are probably closely related to temperature and substrate changes (Harris, 1959; Conover and Corner, 1968; Martin, 1968; Butler et al., 1970).

Excretion variations resulting from organism size relate to the same allometric considerations discussed for respiration. Thus, while absolute excretion (μg-at animal^{-1} day^{-1}) varies directly with size, the weight-specific rate (μg-at mg^{-1} day^{-1}) varies inversely (Corner et al., 1965). Organism size is not specified in the zooplankton compartment of the model, so that these considerations are largely ignored. The approximate effect of juvenile–adult size differences is included, however, and this results in a corresponding increase in the weight specific nitrogen and phosphorus excretion of the juvenile biomass.

In the model, temperature is the driving factor for excretion through the thermal effect on respiration [Eq. (28)] which implicitly represents the gross seasonal changes due to biochemical, physiological and behavioral alterations. To the degree that the simple exponential expression reflects these trends, nutrient excretion does also. This is a section where additional work is especially needed in conjunction with an enhanced respiration formulation.

The chemical form of excreted nutrients has been characterized as primarily inorganic (Pomeroy et al., 1963; Corner and Newell, 1967; Hargrave and Geen, 1968; Peters and Lean, 1973). Organics are also released, and this fraction is a topic of current research, but is assumed here to be less important. Thus, in the model, the daily excretion of nitrogen and phosphorus is added directly to the ambient concentrations of ammonia and phosphate.

Excretion data are frequently related to the oxygen consumption for the same time period. The ratio of these rates depends biochemically upon the substrate being utilized. Generally, for carbohydrate metabolism, the atomic O : N ratio is greater than 17. Mayzaud (1973) suggests O : N > 10 as representative of well-fed zooplankton metabolizing phytoplankton food, although general physiology principles and the work of Conover and Corner (1968) support higher ratios of 16 to 30 or more. Protein metabolism normally results in an O : N around 8, but high nitrogen-base composition of some zooplankton protein may explain O : N values of 4 or lower (Mayzaud, 1973). Under conditions of extreme starvation, even lower ratios may result, perhaps due to traumatic cellular stresses, though these conditions are not documented in nature and the mechanisms are speculative (Mayzaud, 1973). Throughout temporal cycles, zooplankton presumably utilize substrates ranging from freshly ingested, high carbohydrate foods, through stored fat or lipid reserves, into starvation metabolism of proteins. While no storages are allowed to accumulate in the model, the excretion ratios do depend on the food intake versus the metabolic demands.

During periods of phytoplankton abundance, more nitrogen must be retained than carbon in order to produce the relatively protein-rich zooplankton tissue. Some work suggests that the balance may be made up by increased assimilation of N and P over carbon, though the concentration is hard to document due to uncertainty in all methods of assimilation measurement (Mashall and Orr, 1955, 1955a). In the model, this imbalance between food and tissue nutrient ratios decreases the nitrogen release at a given temperature during these periods. When food intake is low, or the food composition approaches that of the zooplankton themselves either by cannibalism or ingestion of bacteria-enriched organic detritus, nutrient excretion is proportionate to respiration in the zooplankton tissue ratio.

For example, consider the model calculations for three situations at the same temperature where respiration is 20% of the body carbon per day (Table 9). In the first, phytoplankton biomass is low and zooplankton ingestion cannot meet the respiratory demands. In the second, respiratory demands are more than met, but on a particulate carbon diet enriched in nutrients by bacterial action to a composition similar to the animals. And in the third, the zooplankton are reproducing and growing on a phytoplankton diet. Maximum nitrogen excretion occurs in Case II (2.8 μg-at N mg C^{-1} day^{-1}), when growth is taking place on a nitrogen-rich diet. This is analogous to exclusively carnivorous feeding. An estimated O:N = 14.9 is lower than the theoretical 17 expected for carbohydrate metabolism. In Case I, excretion is also high, since respiratory demands are being met with protein body tissue (assuming no fat or oil storages), although the small phytoplankton ingestion helps somewhat, reducing N-excretion slightly. In Case III, active growth on the relatively nitrogen-poor food requires retention of nitrogen, thus minimizing excretion. The high estimated O: N is in line with observations for well-fed zooplankton, and even higher ratios result for higher ingestion and growth rates.

Notice that in Cases II and III, where growth is occurring, that nutrient excretion is simply:

(Total assimilated) — (Nutrient necessary to build new zooplankton tissue).

That is, ingested food is effectively being used for all metabolic work—respiration plus growth—though this is true only in the net sense, as protein turnover must constantly take place. In Case I, by contrast, ingestion provides fuel for only a small part of respiration, with zooplankton tissue being consumed for the remainder. The excretion calculation in this case is:

(Total nutrient assimilated) + (Nutrient in respired zooplankton tissue).

The algorithm employing positive or negative change is conveniently identical in all cases, simplifying the computer programming. The following equations are used in the model to calculate nitrogen and phosphorus excretion. For computer efficiency the calculation has been rewritten in a more direct form, but the logic and result is the same as that presented in the tabular example.
Excretion is computed:

$$\begin{aligned} \mathrm{EXN} &= \mathrm{N:C_Z \cdot TOTRSP - XNPZ \cdot TOTRP - XNPCZ \cdot TOTRPM} \\ \mathrm{EXCP} &= \mathrm{P:C_Z \cdot TOTRSP - XPPZ \cdot TOTRP - XPPCZ \cdot TOTRPM} \qquad (31) \end{aligned}$$

Table 9. Sample calculation of nitrogen excretion for 3 nutritional cases. Assume: Zoo biomass = 1 mgC/L; $C:N_{200}$ = 5.95 or N:C = 14 µg-at/mgC; $C:N_{pc}$ = 5.95 or N:C = 14 and $C:N_{ppl}$ = 8.33 or N:C = 10

	Notation	Case I	Case II	Case III	Units
1. Respiration	TOTRSP	0.2	0.2	0.2	mg C·day^{-1}
2. Phyto. ingested	TOTRP	0.05	0.0	0.5	mg C·day^{-1}
" assimilated (0.8)	ASSIM	0.04	0.0	0.4	
3. Particulates ingested	TOTRPM	0.0	0.5	0.0	mg C·day^{-1}
" assimilated (0.8)	ASSIM	0.0	0.4	0.0	
4. N-assim. in food	ASSIM·N:C (2+3)	0.4	5.6	4.0	µg-at N·day^{-1}
5. Net biomass change (growth or resp. death)	URA = ASSIM − TOTRSP	− 0.16	+ 0.2	+ 0.2	mg C·day^{-1}
6. N-required for growth, or lost by resp.	URA·N:C_{zoo}	− 2.24	2.8	2.8	µg-at N·day^{-1}
7. N-excreted	EXN = N-assim (4) − N-metabolized (6)	2.64	2.8	1.2	µg-at N·day^{-1}
8. N exc:C resp.	EXN (7)/TOTRSP (1)	13.36	14.0	6.0	µg-at N·mg C^{-1}
9. C:N (by atoms)	TOTRSP/EXN	6.25	5.95	13.89	µg-at C·µg-at N^{-1}
10. O:N RQ = 1.0 (CHO)	(9)·2.0	12.50	11.90	27.78	µg-at O·µg-at N^{-1}
RQ = 0.8 (Prot.)	(9)·2.5	15.63	14.88	34.73	
11. Fraction of body N excreted		19.1	20.0	8.6	%

where:

$$TOTRSP = RESP + RESPJ = \text{combined adult and juvenile respiration } (\text{mgC l}^{-1}\ \text{day}^{-1}),$$

$$N:C_Z,\ N:C_P,\ N:C_{PC} = \text{nitrogen-to-carbon ratios for zooplankton, phytoplankton, and particulate carbon},$$

$$P:C_Z,\ P:C_P,\ P:C_{PC} = \text{phosphorus-to-carbon ratios},$$

$$TOTRP = \text{combined adult and juvenile ration of phytoplankton},$$

$$TOTRPM = \text{combined ration of particulate matter},$$

XNPZ, XNPCZ, XPPZ, XPPCZ = correction factors for unequal nutrient rations assimilated vs. zooplankton tissue, i.e., assimilation ×(difference in ratios). Thus:

$$XNPZ = XASSIM \cdot (N:C_Z - N:C_P).$$

The example presented in Table 9 represents the excretion calculation for an overall zooplankton C:N=6. If the contention of Mayzaud (1973) is correct, and the metabolic substrate utilized by unfed zooplankton has a C:N of 2–3, the equation predicts C:N and O:N ratios in agreement with his observations. In the model, however, it would be inappropriate to use such low ratios, since the average body composition is being represented. As in the phytoplankton nutrient formulation, cellular metabolic pools of a specific composition are beyond the scope of this model. Nevertheless, even across the reported ranges of average nutrient compositions, the simple scheme presented here results in excretion rates and ratios in good agreement with most laboratory and field observations, as well as with theoretical considerations.

5.6 Reproduction

Energy or material budgets of animal metabolism generally balance assimilation against growth, reproduction, and respiration. This assumes that all processes not directly manifested as growth or reproduction, such as behavioral activities of various types, are effectively grouped together as unproductive respiratory expenditures, at least for some purposes. In this model, the generalized zooplankton compartment and respiration formulation make this assumption appropriate. Further, if adults are defined as having completed growth, all unrespired assimilation (URA) goes directly to reproduction. Egg production by adult zooplankton may be represented simply by this elementary mass balance. Thus, if:

$$RESP = \text{adult respiration } (\text{mgC l}^{-1}\ \text{day}^{-1})$$

$$ASSIM = \text{adult assimilation } (\text{mgC l}^{-1}\ \text{day}^{-1})$$

then:

$$URA = ASSIM - RESP\,. \tag{32}$$

When URA is negative, starvation results in a decrease in the adult zooplankton carbon biomass. When URA is positive, eggs equal to the difference are produced and dispersed into the water. While some species of copepods retain their eggs for a time in a brood sac, *Acartia* species that frequently dominate the bay copepod population release their eggs directly.

Growth of juveniles to adulthood is often represented by a sigmoid curve (e.g., von Bertalanffy, 1957). Winberg (1971, pp. 113–128) has demonstrated that if egg production is considered, the exponential or, more accurately, parabolic curve of early growth through time flattens much more slowly to a plateau. This observation supports the basic assertion that both reproduction and growth may be considered metabolically similar processes in that each involves a more-or-less direct conversion into zooplankton tissue of the assimilated ingestion which exceeds respiration.

A number of assumptions are implicit in this formulation. First, the nutritional enrichment producing the yolk which supports the developing embryo during the incubation period is not considered. The model assumes a single carbon : nutrient composition for zooplankton adult, juvenile, and egg tissue. While this is not an accurate assumption, maintenance of such detailed metabolic variations is beyond the scope of a generalized ecosystem model. Second, the coupling of reproduction to respiration and ingestion necessarily invokes all the assumptions of those formulations. The generality of the respiration expression with respect to the increased metabolism associated with reproduction has already been mentioned, and the role of temperature in increasing the biological rate processes of respiration and ingestion is obviously important in reproduction as well. The hyperbolic food-density dependence of egg production in zooplankton has been documented. Field observations on fresh-water organisms (Edmondson, 1961) suggest that the reproductive rate increases to a plateau with increasing ambient algal concentrations, and direct laboratory experiments with marine species (Marshall and Orr, 1952; Heinle et al., 1973) depict a clearly hyperbolic response of egg production with food availability. Since many of the observations which support the "Ivlev" pattern were on adult animals that were presumably reproductively mature, the conclusion that a similar fecundity response will result is reasonable.

Finally, the reproduction formulation presented here assumes immediate conversion and release of eggs. This is not unreasonable, especially within the one day time-step of the numerical calculations. Marshall and Orr (1952) demonstrated that egg production of adult female *Calanus* from nature was measurably enhanced the day following the initial feeding after 15 days of starvation (Marshall and Orr, 1952). However, they also commented that a short delay of two to five days may separate feeding and egg production, depending on the sexual maturity of the female.

5.7 Egg Hatching Time

For most species in the bay, and therefore in the model, eggs released by adult zooplankton drift in the water column during a period of incubation before hatching. Again, as a metabolic process the incubation would be expected to be an exponential function of temperature. In this case, thermal enhancement of the development process will speed hatching, thus an inverse logarithmic expression for hatching time at increasing temperatures is expected. A number of workers have documented this trend for a variety of species (Edmondson et al., 1962; Comita and Comita, 1966; Burgis, 1970; Landry, 1975). McLaren (1965, 1966; McLaren et al., 1969) has presented data for a number of marine and estuarine copepod species which permit a regression analysis. Additional data for copepods from the Chesapeake Bay region (Heinle, 1969a) fall nicely along the same trend (Fig. 29). McLaren et al. (1969) adopted an equation reportedly appropriate for relating physiological rates to temperature,

$$H = a(\text{Temp} - \propto)^{-2.05}$$

where: H = hatching time (days),
$a, \propto$ = constants.

Specific species variations were demonstrated in the trends, with a clear correlation of one equation parameter ($\propto$) with estimates of average environmental temperature, suggesting measurable adaptation by the tested species. While the predictions of this equation are good, there is also satisfactory agreement with the more conventional Q_{10} relationship, which was chosen for use in the model.

Combining all the reported data (Fig. 29), the regression equation for hatching time (H, days) was:

$$H = 12.0\,e^{-0.11 \cdot \text{Temp}}. \tag{33}$$

Thus, at a temperature of 0° C, eggs are delayed 12 days during incubation. While this is short for some species, it is appropriate as a seasonally averaged trend taking into account the acclimation to environmental temperatures demonstrated by McLaren et al. (1969). The close agreement with the 0° C observation for the typical winter inhabitant of Narragansett Bay, *A. clausi*, is encouraging (Fig. 29). In the model, the hatching time for the eggs released on a given day (+URA) is determined from the temperature for the specific region of the bay on the day of release. Although spatial and temporal changes in temperature may result in an altered hatching time in nature, for the short time span involved these effects are certainly small, and they are omitted from the model. During the incubation period, the egg biomass circulates around the model bay, and is subject to the combined pressure of adult cannibalism and carnivorous zooplankton predation. No other natural mortality is included, as egg viability is probably high (Paffenhöfer, 1970), but the combined losses to flushing and predation result in only a fraction of the eggs finally hatching in the bay, as might be expected.

5.8 Juvenile Development

Upon hatching from the egg compartment, juvenile zooplankton begin development to sexual maturity. Some aspects of juvenile metabolism, including respiration, excretion, assimilation, and the theoretical background to food

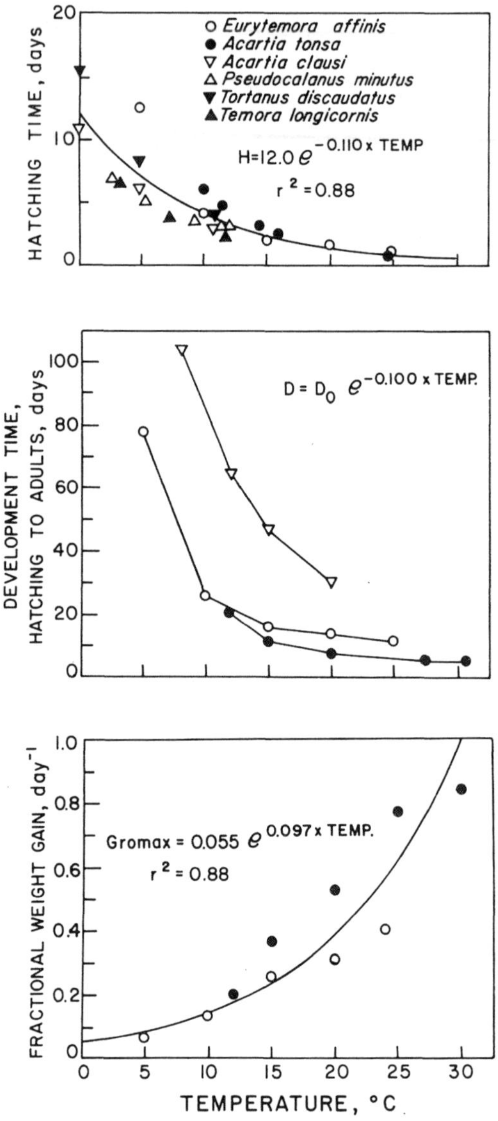

Fig. 29. Hatching time of eggs *(top)*, development time to maturity *(middle)*, and maximum daily growth rate of juveniles *(bottom)* for representative estuarine copepods. Equations fit by least squares regressions. Data from McLaren et al. (1969) and Heinle (1966). (From Nixon and Kremer, 1977)

limitation are treated very similarly to the adults and have already been discussed. Because the focus of the juvenile section is growth, in contrast to adult reproduction, a slightly different approach is taken. For adults, ration and respiration are calculated explicitly, with the difference representing potential egg production. For juveniles, growth and respiration are explicit, from which the necessary ration is estimated. This approach was chosen because the juvenile development literature concentrates on such growth estimates, with relatively less direct evidence on rations. It should be mentioned that the alternative approach was also modeled for comparison, and results were quite similar.

Juvenile development may be viewed as the combination of two aspects: a growth rate, and a specified development time. Both of these are strongly temperature dependent, and both may reflect the additional influence of food availability. While other considerations undoubtedly may also effect the growth process, the model formulation concentrates on these two primary interactions.

Briefly, the maximum daily growth rate is estimated along with respiration as a function of temperature. With a correction for assimilation efficiency, a ration required to support this growth may be determined. Food limitation, based on the Ivlev formulation, reduces this preferred ration to an appropriate level. Both phytoplankton and detritus compose the available food resource, as cannibalism is assumed not to occur by juveniles. Development time is estimated taking into account the anticipated changes in temperature that may result during the sometimes prolonged period. Scarcity of food may slow actual development, and predation by adults and carnivores decreases the stock during growth. At the end of the developmental period, a biomass of grown juveniles moves into the adult compartment where reproduction may occur, completing the life cycle.

5.8.1 Growth Rate of Juveniles

The formulation chosen to express the temperature dependence of growth rate is the familiar exponential Q_{10} relation.

$$\mathrm{GROMAX} = \mathrm{ZJGMX}_0\, e^{0.1 \cdot \mathrm{Temp}} \tag{34}$$

where: ZJGMX_0 = maximum juvenile instantaneous growth rate at 0° C (day^{-1}).

In theory, GROMAX represents the maximum instantaneous rate of juvenile growth (day^{-1}), the rate achieved at a given temperature when food is not limiting. In fact, this ideal number is unavailable, since neither laboratory nor natural conditions may be assumed to represent this situation. Evidence for the general temperature effect is readily available (see Mullin and Brooks, 1970). Heinle (1969a) reported measurements for two estuarine copepod species in Maryland for conditions of rapid growth when, presumably, there was little food limitation (Fig. 29). For each species alone there was some indication of upper thermal limits, but if the combined data set is considered representative of a seasonally acclimating community, the exponential regression is strong, and the zero-degree intercept of 5.5% daily is a reasonable value. The temperature coefficient of about $0.1°\,\mathrm{C}^{-1}$ represents a $Q_{10} = 2.7$, and is also reasonable considering the adaptive value of enhanced developmental rates.

5.8.2 Development Time

Like hatching time, the duration of development decreases as the rate increases. Thus, an inverse exponential equation is suggested. As with the closely related GROMAX, the calculated development time is meant to represent the minimum under ideal conditions. Observational data rarely approach this, and some flexibility must be retained in the model formulation. Selected data are presented in Figure 29, and fitted to the model equation:

$$D = D_0 \,\mathrm{e}^{-0.1 \cdot \mathrm{Temp}}. \tag{35}$$

The temperature coefficient proves to be remarkably similar in the two widely different sets of measurements. Heinle's (1969a) data for the Chesapeake suggest $-0.099°\,\mathrm{C}^{-1}$ for *E. affinis*, and data of Greze and Baldina (1964) working with *A. clausi* from the Black Sea fit a coefficient of $-0.101°\,\mathrm{C}^{-1}$. The zero degree intercepts (D_0) are very different, however, being 92 days for *E. affinis* and 225 days for *A. clausi*. These data indicate that fundamentally the same process is functioning, although, again, local species succession and acclimation may change absolute values. Accordingly, the temperature coefficient is fixed in the model as $-0.1°\,\mathrm{C}^{-1}$, while the zero degree development time, D_0 (referred to in the actual program as JZMAX, the maximum number of juvenile compartments) is a variable to be specified at the time of the simulation. Because development may span many weeks, a correction based on the anticipated temperature change is also necessary. However, since this correction involves some intricate computer programming, it will be discussed in Chapter 7.

An interesting analysis may be developed relating the necessary interdependence of growth rate and development time. The pattern of growth may be represented by the standard instantaneous growth equation:

$$Z = Z_0 \,\mathrm{e}^{G},$$

where:

$$G = \mathrm{GROMAX} \cdot D.$$

If the ratio of adult to juvenile size (Z/Z_0) is to remain constant, the exponent G must not vary with temperature. In other words, the temperature coefficients for GROMAX [Eq. (34)] and D [Eq. (35)] must be approximately equal and of opposite sign, with small inequalities resulting in a temperature dependence of adult size. This temperature effect is well known, with larger adults generally typical of colder waters (Conover, 1956; Heinle, 1969a), suggesting that the equality is not precise. In a model where body size was simulated, choice of slightly unequal exponents would result in seasonal variations in size. However, for this model this scale of detail was not justified. The good agreement in the fitted exponents of GROMAX and D is ultimately indicative of the fact that the relative egg-to-adult sizes are comparable, both for the different species, and for various temperatures. Nevertheless, this agreement for such disparate data is encouraging.

5.8.3 Food Limitation

The rationale and interpretations of the commonly observed hyperbolic ingestion response to variations in food concentrations have already been discussed. An identical formulation is used in the juvenile compartment to govern the actual realized growth and development time through food limitation. Reviewing briefly, Ivlev's (1945) original Eq. (18) may be rewritten in a normalized form to calculate a unitless fraction representing the degree of food limitation XRTNJ [see Eq. (24)]. The ration of the zooplankton juveniles is projected from the weight-specific growth and the respiration rates, GROMAX and XRESPJ ($\text{mgC mgC}^{-1}\,\text{day}^{-1}$). Multiplication by the food limitation term XRTNJ reduces the estimate, which is then used to calculate a filtering rate appropriate to the instantaneous rate integrations. In the equations of the model:

$$\text{XRESPJ} = \text{RJ:RA} \cdot \text{XRESP} \qquad \text{[see Eqs. (28) and (30)]}$$

$$\text{XRTNJ} = 1.0 - e^{-\text{IVLEVK} \cdot \text{PCMG}} \qquad \text{[see Eq. (24)]}$$

$$\text{RTNJ}_{\text{max}} = \text{ZJTOT} \cdot (\text{GROMAX} + \text{XRESPJ})/\text{XASSIM}$$

$$\text{RTNJ} = \text{RTNJ}_{\text{max}} \cdot \text{XRTNJ} \qquad \text{(estimated)}$$

$$\text{RLF} = \text{RTNJ}/\text{PCMG}$$

[simultaneously evaluate rates and compute rations; see Eq. (49)]

$$\text{RTNJ} = \text{RTNJP} + \text{RTNJPM} \qquad \text{(realized)}$$

$$\text{GRO} = \text{XASSIM} \cdot \text{RTNJ}/\text{ZJTOT} - \text{XRESPJ}$$

where:

PCMG = available food concentration (mgC l^{-1}),

ZJTOT = biomass of juveniles (mgC l^{-1}),

XASSIM = fraction of ingestion assimilated,

RLF = filtering rate based on estimated ration and food concentration ($\text{l l}^{-1}\,\text{day}^{-1}$),

RTNJP and RTNJPM = ration of phytoplankton and particulate matter consumed by juveniles (mgC day^{-1}),

GRO = final realized juvenile growth.

The limitation fraction XRTNJ is also used to delay development. The algorithm will be presented later in Chapter 7, but the underlying assumption is that development on a day when XRTNJ = 0.75 is equivalent to only three-quarters of a day of optimum development. That is, while the ingested ration would only be sufficient to promote 75% of the maximum daily growth increment, the development time would be commensurately lengthened by 25%. In the program, of course, these calculations are all made daily in continually changing ambient conditions of temperature and phytoplankton concentration.

Overall, the zooplankton juvenile development formulation results in realized generation times that agree well with published estimates. Jeffries (1962) reviewed data for both *Acartia* species suggesting natural developments of about two months for *A. clausi* in colder waters and one month for *A. tonsa*, the summer species. And microcosm studies found *A. tonsa* developments ranging from 24–31 days (Raymont and Miller, 1962). Observations for *A. tonsa* in laboratory conditions which might be expected to represent well-fed optimal rates range from 20 days (Sosnowski, U. S. Environmental Protection Agency, Narragansett, personal communication) to 25 days (E. Hulsizer, URI Oceanography School, personal communication) at 20° C. With an average $D_0 = 150$ days, the model Eq. (35) agrees with these laboratory studies.

With the food limitation correction, the estimates originally based on temperature are lengthened, resulting in realized developments of about one month, closely paralleling most of the natural estimates (Jeffries, 1962). One reported estimate departs markedly from this range, however. Heinle (1966) observed *A. tonsa* to develop in 6–9 days in the Patuxent estuary in Maryland. These rates have been repeatedly confirmed by Heinle, and recent evidence from his laboratory at Solomon's Island, Md., suggests that low salinities may somehow relate to this atypical finding. This shortening of development in fresher waters is currently being investigated by Dr. Heinle, and would indicate that estuarine modeling may not always conveniently ignore salinity effects even on strongly euryhaline species.

6. Additional Compartments

6.1 Carnivorous Zooplankton

Larval fish and the ctenophore, *Mnemiopsis leidyi*, are among the most important predators on the smaller zooplankton in Narragansett Bay. The biomass of both populations is input to the model by forcing functions that approximate extensive field observations in the bay of the spatial and temporal distributions of ctenophores (P. Kremer and Nixon, 1976) and fish larvae (Matthiessen, 1974). Figures 30 and 31 present the patterns of seasonal abundance used in the model.

The importance of chaetognath predation was considered using the biomass data of Martin (1965) and a feeding potential based on the work of Sameoto (1972). Except for certain parts of the bay for brief times during the year, their estimated role was small, so predation by chaetognaths was not included in the model.

Laboratory data on feeding by ctenophores (P. Kremer, 1975) suggest that they are passive filterers, and that a constant volume-swept-clear for each unit of biomass is reasonable as a first approximation. At the temperature of peak abundance, a rate of 0.1 liters mg^{-1} day^{-1} was chosen as appropriate for an average-size organism. Empirical excretion rates are also applied to the biomass estimates to include the ctenophores in nutrient regeneration. Phosphate, silicate, nitrite and nitrate excretion were low, and therefore are not included, but ammonia excretion (CTEXN, μg-at l^{-1} day^{-1}) is specified as a function of temperature (P. Kremer, 1975a):

$$\mathrm{CTEXN} = \mathrm{CTEN} \cdot 0.0014\, e^{(0.13 \cdot \mathrm{Temp})}$$

where: CTEN = biomass, mg dry weight/l.

The literature on larval fish feeding is limited, although extensive aquaculture research with juvenile fish is useful. In general, the formulation includes determination of a preferred feeding rate as a function of temperature which is adjusted for limited food availability by the hyperbolic relationship originally suggested by Ivlev (1945) and used earlier in the zooplankton formulation. Since the larvae are primarily if not exclusively visual feeders (Blaxter, 1965), the final daily instantaneous feeding rate is reduced by the photoperiod.

The choice of the preferred maximum ration is based on a variety of work with different species, some of which was not reported directly as ration and which required conversion assumptions in estimation. For example, using stomach weights and gut evacuation times for large mouth bass larvae (G. C. Laurence, Natl.

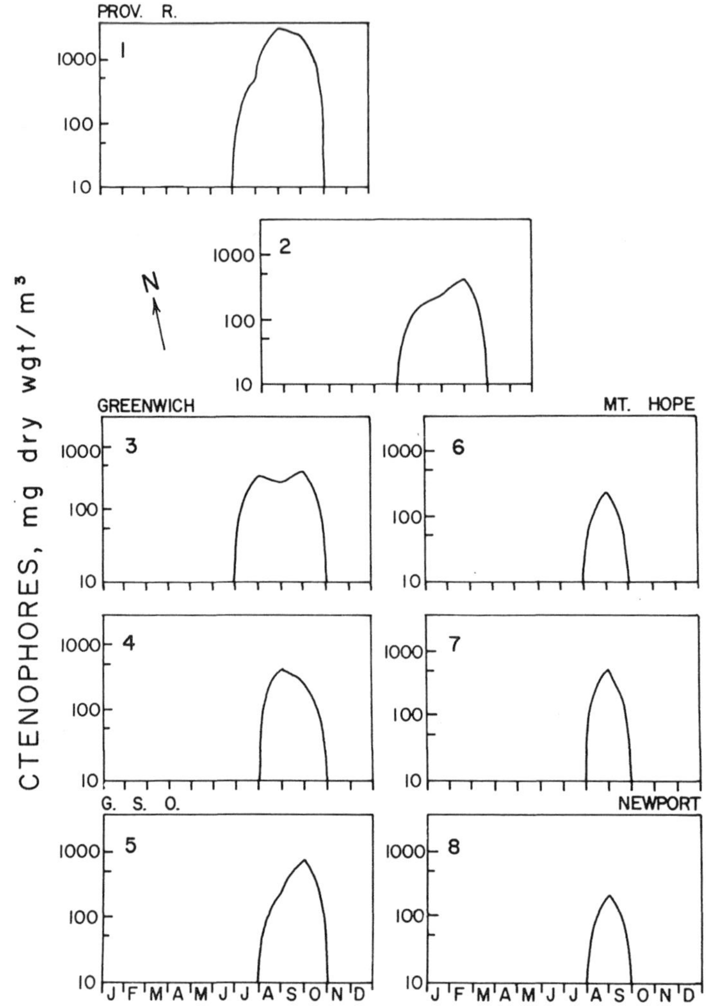

Fig. 30. Seasonal patterns of biomass (μg dry wt/l) of the ctenophore, *Mnemiopsis leidyi*, used in the eight spatial elements of the bay model. Based on data of P. Kremer and Nixon (1976) for Narragansett Bay, 1972

Mar. Fish. Serv., Narragansett, R. I., personal communication), it appears that the daily ration may amount to 20–50% of the body weight. Recent work on specific growth rates of winter flounder (Laurence, 1975) led to estimates of 10–30% for 0.5–3.0 mg dry wt larvae tested at 5 and 8° C. Other published estimates fall into the range of 40–60%/day (Sorokin and Panov, 1965; Pandian, 1967). Although some estimates are higher (Blaxter, 1965; Ghittino, 1972) and lower (Parsons and LeBrasseur, 1968), a choice of a zero-degree maximum ration of 25% of the body weight per day seems reasonable. An arbitrary $Q_{10}=2$ is used to adjust the rate for temperature as suggested by the fish-culture guidelines of Ghittino (1972). To calculate the carbon requirement represented by this ration, an estimate of the size

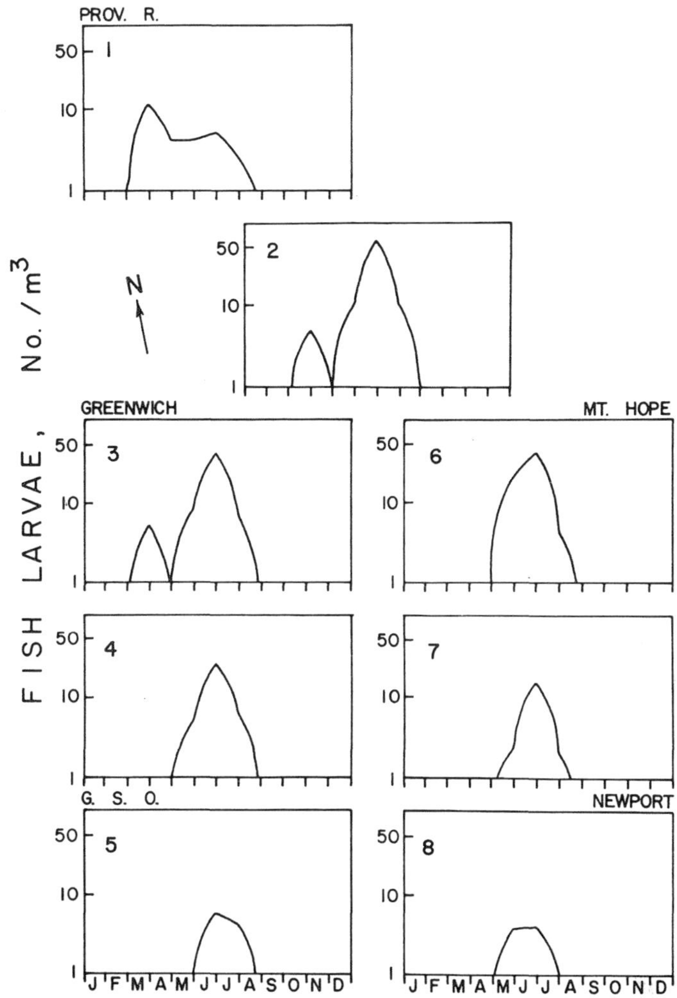

Fig. 31. Seasonal patterns of abundance (no./m^3) of fish larvae used in the eight spatial elements of the bay model. Based on data of Matthiessen (1974) for Narragansett Bay, 1972–1973

and carbon content of the larvae was necessary. A ration value of 0.15 mgC per larvae per day was chosen based on the size range of winter flounder, haddock, tautog and menhaden larvae (Laurence, unpublished; Bigalow and Schroeder, 1953) and an assumption that 50% of the dry weight is carbon.

Evidence supporting the general hyperbolic relationship suggested by Ivlev (1945) between food abundance and ration consumed has been presented for fresh and salt water species (Laurence, personal communication; Parsons and LeBrasseur, 1968). Estimation of the exponent in the equation, however, is difficult. A value of 30 l/mgC was chosen based on only semiquantitative observations. This value will allow almost the maximum ration at the higher zooplankton concentrations in the bay (95% of maximum at 0.10 mgC/l). While there is considerable uncertainty in all parameters of this formulation, the basic form is well supported. Future expansion of this compartment to include growth of the fish larvae in a mechanistic way may allow the consistency of the choices to be evaluated.

The equations for the carnivorous zooplankton feeding in each liter of the bay may be summarized as follows:

$$\mathrm{FRMX} = \mathrm{FRMX}_0 \cdot e^{(\mathrm{QFRMX} \cdot \mathrm{TEMP})}$$

$$\mathrm{FSHRTN} = \mathrm{FRMX} \cdot \mathrm{PHOTPD} \cdot \mathrm{FISH} \cdot 0.001$$

$$\mathrm{FRTN} = \mathrm{FSHRTN} \cdot (1 - e^{(-\mathrm{IVFSH} \cdot \mathrm{ALLZ})})$$

$$\mathrm{FFISH} = \mathrm{FRTN}/\mathrm{ALLZ}$$

$$\mathrm{FCTEN} = \mathrm{CTFLT} \cdot \mathrm{CTEN}$$

$$\mathrm{FCARN} = \mathrm{FFISH} + \mathrm{FCTEN}$$

where: FRMX_0 = maximum daily ration at 0° C (mgC larvae^{-1} day^{-1}),

QFRMX = temperature exponent, for Q_{10} of 2.0=0.069,

FRMX = temperature corrected maximum ration (mgC larvae^{-1} day^{-1}),

CTEN, FISH = biomass of ctenophores (mg dw l^{-1}) and larval fish (No. m^{-3}, 0.001 converts to No. l^{-1}),

PHOTPD = daylight fraction for visual feeding,

FSHRTN = preferred ration of the fish larvae (mgC day^{-1}),

IVFSH = hyperbolic limitation parameter, see Eq. (18) (l mgC^{-1}),

ALLZ = total available food: zooplankton adults, juveniles, and eggs (mgC/l),

FRTN = projected ration of fish larvae (mgC day^{-1}),

CTFLT = weight-specific ctenophore filtering rate (l day^{-1} mg^{-1}),

FCTEN = instantaneous filtering rate of ctenophores (l day^{-1}),

FFISH = instantaneous feeding rate of fish larvae (l day^{-1}).

As with copepod grazing, the final realized rations are calculated by an explicit evaluation of the instantaneous feeding rates (see Chap. 7).

6.2 Adult Fish and Higher Trophic Levels

Recent experiments suggest that the predation of adult Atlantic Menhaden, *B. tyrannus*, on copepods may be of seasonal importance in the upper reaches of Narragansett Bay. The presence of a lower size threshold for feeding (Durbin and Durbin, 1975) and the clear-cut dominance of small flagellates in the Providence River region of the bay during the summer (E. Durbin et al., 1975) indicate that this

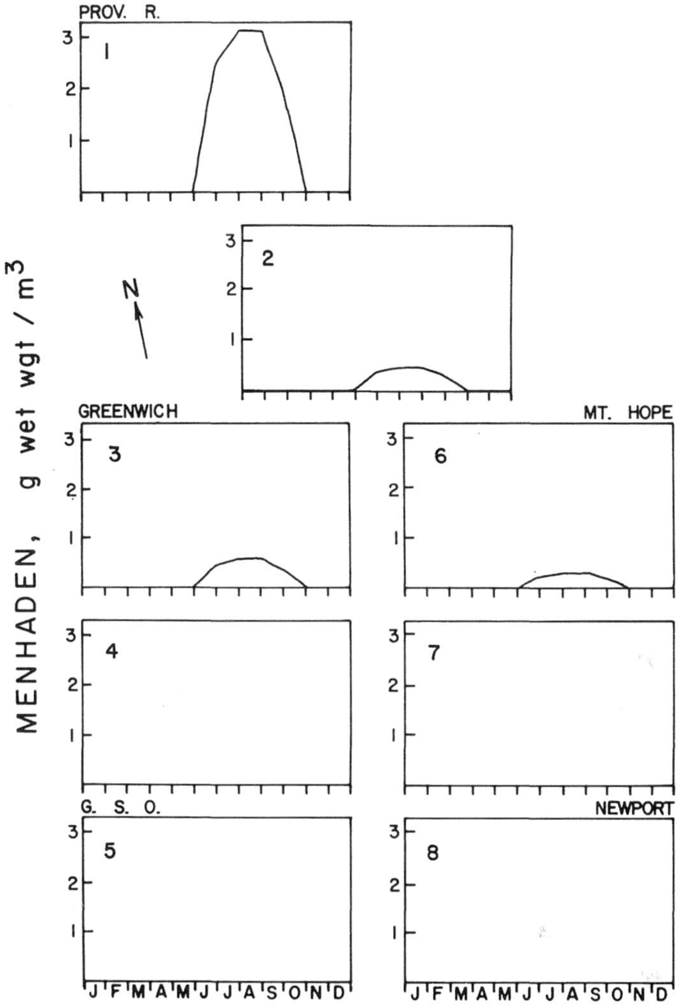

Fig. 32. Seasonal patterns of biomass (g wet wt/m^3) of Atlantic Menhaden, *Brevoortia tyrannus*, used in the eight spatial elements of the bay model. Based on typical spotter estimates and commercial catches compiled by A. Durbin (1976) for Narragansett Bay

carnivorous diet may be more important than the more traditional herbivorous mode. The effects of this additional predator have been included in the model in a manner similar to that of fish larvae.

Biomass estimates for the menhaden have been compiled from Rhode Island commercial landings and airplane spotter estimates (A. Durbin, 1976). Catch statistics suggest a seasonal pattern of abundance from June through October with peak biomasses during August. Spotters estimate the maximum abundance to be around 15 million pounds wet weight, most of which is concentrated in the Providence River and upper bay regions. Accordingly, an appropriate pattern of monthly biomass is forced in the model apportioning 60% of the total wet weight in

Element 1 (Providence River), 20% in Element 2, and 10% each in Elements 3 and 6 (Fig. 32).

Using measured respiration and excretion rates (A. Durbin, 1976), and a representative growth rate, a crude estimate of metabolic requirement is 6.25 kcal per pound wet weight of fish per day. Assuming 4.5 kcal/g dry weight and 35% carbon for zooplankton, a requirement of 0.5 g C per pound per day may be calculated. The familiar Ivlev (1945) expression for food-density dependence was used, with the inclusion of a concentration threshold [Eq. (25)] since active feeding appears to stop below a density of around 12 copepods per liter (Durbin and Durbin, 1975), or approximately 0.03 mg C/l. The choice of the exponent k in the Ivlev equation was arbitrary. The value of 33 l/mg C allows 90% of the maximum ration at a zooplankton concentration of 0.1 mg C/l. These parameter choices result in a formulation which switches rapidly to the maximum feeding rate at concentrations above the threshold, in agreement with experimental observations cited earlier. As with the fish larvae, the estimated ration is converted to an average daily filtering rate which, in combination with ctenophore and fish larvae, decreases zooplankton adults, juveniles and eggs.

The effects of other adult fish and higher trophic levels are omitted from the model. Though indirect controlling factors are often difficult to identify, it is believed unlikely that this predation exerts significant influences on the planktonic system. Certainly not enough is presently known to permit inclusion in the present modeling attempt, nor is detailed treatment of all trophic levels technically practical in view of the analytical scope of the planktonic model compartments.

6.3 The Benthos

The role of the benthos in the model is in two parts: phytoplankton grazing, and nutrient regeneration. The grazing formulation is based on the large and widespread populations of the hard clam, *Mercenaria mercenaria*, found throughout most of the bay, and has been modified from an earlier conceptual model of clams and the clam fishery developed by Lampe and Nixon (1970). Unfortunately, there are few data available on the dynamics of this, or other, estuarine benthic communities, and an active program to measure feeding, respiration and nutrient regeneration rates in situ is now underway. Because of uncertainty in the literature, considerable flexibility has been built into the clam grazing formulation. Simply, a daily clearing rate for the clam population is determined from pumping rate and duration projections as a function of temperature. Nutrient regeneration with temperature is according to in situ flux measurements made in Narragansett Bay by Nixon et al. (1976).

6.3.1 Grazing by Clams

Russell (1972) carried out a clam population survey of the West Passage near the Jamestown Bridge. He estimated 35–40 bushels (80 pounds each) per acre prior to the commercial dredging season. Although the subsequent harvest lowered the population to 14 bushels/acre, it can be assumed that some replacement of the loss

Table 10. Clam population estimates (number/m^2) used in 8 spatial elements of the model (Fig. 1). Values based on findings by Russell (1972) for element 4 of 35 bushels/acre, and prorated around the bay assuming 300 70 mm clams/bushel

Spatial element	Relative abundance to element 4	Prorated population no./m^2
1	1.25	3.23
2	1.20	3.10
3	1.10	2.84
4	1.00	2.58
5	0.90	2.32
6	1.10	2.84
7	0.75	1.94
8	0.25	0.65

occurred, and the 35 bushels/acre value was chosen as a reasonable stock estimate. The wide confidence limits for the methods preclude a closer interpretation of the population size. While these estimates by Russell are among the best available, they are quite limited spatially. To provide baywide estimates, relative spatial abundances shown by earlier studies (R. I. Department of Natural Resources Shellfish Survey Leaflets; Saila et al., 1967) were used to prorate Russell's relatively detailed estimate around the bay. An average size of 70 mm (shell diameter), corresponding to about 300 clams/bushel, was chosen as representative of the "cherrystone-chowder" size distribution reported (R.I. Department of Natural Resources). The population estimates are indicated in Table 10.

A pumping rate of about 5.0 l/h for a 70 mm clam was chosen based on general agreement of a variety of data. Walne (1972) reported 67–73 ml/min (4.2 l/h) for 60–70 mm clams at 20–21°C, and Hamwi and Haskins (1969) show rates of from 1–10 l/h for clams of unspecified size. Rice and Smith (1958), again for unspecified sizes, demonstrated 3–6 l/h clearing rates for clams at 23° C with a suspension of diatoms, although lower rates of 2 l/h or less occurred with small green algae. Direct measurements of flow for various sizes of clams were made by Coughlan and Ansell (1964). Using their regressions, "typical" and "maximum" pumping rates for a 70 mm clam are 5.7 and 6.8 l/h, respectively. Another approach measured the weight-specific oxygen consumption of clams as a function of size, from which estimates of pumping rate were made by assuming an oxygen utilization efficiency (Loveland and Chu, 1969). Although the utilization coefficient was chosen so as to maximize the agreement with the measurements of Coughlan and Ansell, the slopes of the temperature effect were virtually identical, adding credence to both sets of observations.

A response such as pumping rate might be expected to show an exponential temperature relationship with an appropriate Q_{10}. However, no experimental verification of this could be found. Loosanoff (1939) demonstrated essentially a

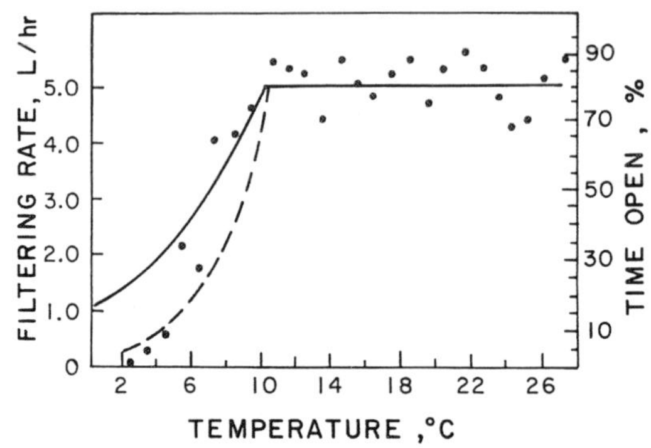

Fig. 33. Filtering rate of clams as a function of temperature. Data from Loosanoff (1939) for fraction of the day clams were open and presumably pumping. The model formulation calculates filtering rate directly. *Broken line* demonstrates a pattern which agrees qualitatively with the observations of Loosanoff. *Solid line* was used in the model, defining a filtering rate of 1.0 l/h at 0° C which rises rapidly to a maximum of 5.0 l/h at 10° C

temperature threshold for the fraction of the day when the valves are open, and the clams are presumably pumping (Fig. 33). Initially, clam grazing was formulated to allow pumping at the 5 l/h rate above the 7° C critical temperature. To give the formulation more generality, however, a pumping rate as an exponential function of temperature was proposed, in combination with a lower temperature threshold and an arbitrary upper limit to pumping rate.

If (TEMP > TLOLIM):

$$\mathrm{VFILT} = \mathrm{VFILT}_0 \cdot e^{(\mathrm{QVFILT} \cdot \mathrm{TEMP})} \tag{36}$$

$$\mathrm{IF\ (VFILT > VMAX)\ VFILT = VMAX.}$$

Thus,the threshold pattern of Loosanoff (1939) may be approximated by a large temperature effect above a 7° C threshold ($Q_{10} \gg 2$), reaching a plateau at the upper rate limit of 5 l/h at about 10° C (Fig. 33). Alternatively, a more conventional Q_{10} may be used which produces the 5 l/h rate at a single-specified temperature, with the threshold and upper limit chosen to have no effect. In any case, it is assumed that the pumping rate represents the volume effectively cleared of phytoplankton with 100% filtering efficiency. This does not, of course, suggest 100% assimilation by the clams at all phytoplankton concentrations. The deposition of pseudofeces at high food concentrations is well known (see Rice and Smith, 1958), and a model of the clam dynamics would have to include density-dependent assimilation. To the plankton community, however, the loss is the same, and the independence of filtering rate and food density is reasonable (Rice and Smith, 1958). As with other grazing rates, the instantaneous filtering rate is combined with the other phytoplankton growth and loss terms, and the actual phytoplankton loss to the benthos is determined by an explicit evaluation of the simultaneous rates (see Chap. 7).

It is interesting to consider the nutritional balance of the clams, even though no mechanistic formulations of these feedbacks are presently included in the model. Loveland and Chu's (1969) respiration and pumping rate regression equations can be interrelated mathematically to represent oxygen consumption (R, ml O_2/h) as a function of pumping rate (PR, l/h). At 25° C:

$$\log \Delta O_2 (\mathrm{ml}\, O_2\, g^{-1}\, h^{-1}) = -0.344 \log \mathrm{wt}\,(g) - 1.023$$
$$\log \mathrm{PR}\, (l\, h^{-1}) = 0.656 \log \mathrm{wt}\,(g) - 0.569\,.$$

Equating these two expressions:

$$\log R = \log \mathrm{PR} - 0.454$$
$$R = 0.352\, \mathrm{PR}\,. \tag{37}$$

This relationship was also studied by Hamwi and Haskins (1969), who reported for 24° C:

$$R = 0.894\, \mathrm{PR} - 0.748\,. \tag{38}$$

This linear function was proposed for pumping rates between 1 and 12 l/h, although Verduin (1969) pointed out that a hyperbolic response is more reasonable.

The two Eqs. (37) and (38) differ by more than a factor of 2 at 5 l/h. The consistency between Loveland and Chu's results and Coughlan and Ansell's carefully documented data lead us to accept their results. Assuming 1 g C is respired for 2 g oxygen (RQ = 0.75), an estimate of the carbon requirement for pumping (CREQ, mgC h^{-1} $clam^{-1}$) is possible:

$$\mathrm{CREQ} = 0.252 \cdot \mathrm{PR}\,. \tag{39}$$

While this computation does not affect the model results, it is interesting to compare the carbon gained with the respiratory requirement. Lampe and Nixon (1970), in a preliminary clam model, proposed a "cost-benefit" scheme where the clams would only pump when enough food was present to meet their respiratory demands. This constraint was eliminated from the present model because the condition was rarely met for reasonable phytoplankton concentrations, even using the lower of the two reported respiration equations. The inconsistency of the pumping rate and respiration demand formulations may suggest problems in the expressions themselves. However, a reasonable alternative explanation is that a substantial part of the metabolic demands of the clams in nature is met by foods other than phytoplankton, such as organic detritus. This potential role of nonliving food sources is also suggested by the model for the copepods, especially at critical times of the year. More information about the magnitudes and utilization of detritus in the bay would appear to be an important contribution to our understanding of the bay system.

6.3.2 Nutrient Regeneration

The regeneration of nutrients by the sediments is based on recent work in Narragansett Bay (Hale, 1974; Nixon et al., 1976, and in preparation). An on-going program has measured in situ fluxes using bottom chambers throughout the year at a number of stations around the bay. These measurements may be used to relate the nutrient dynamics of the total benthic community to temperature. Thus, while the infauna and bacteria are not explicitly treated, their combined effects are included in the model benthos. Large variability in the replicate field flux measurements (coefficient of variation = 25%) was nevertheless substantially less than the variability of standard-grab biomass estimates (coefficient of variation = 75%). Moreover, these studies have included measurements of the responses of three representative community types, including amphipod mats, a polychaete-bivalve assemblage, and a clam bed, yet differences in the fluxes as a function of temperature were usually not large. Thus, although small scale patchiness certainly exists, the functional properties of the systems appear to be less variable.

For the purposes of the model, uniform area-based fluxes as functions only of temperature were assumed for the benthos throughout the bay. This assumption is perhaps weakest in the Providence River, where heavy sediment and effluent loads may drastically alter the characteristics of the bottom. Phosphate-phosphorus, silicate-silica, and ammonia-nitrogen are regenerated according to the relationships shown in Figure 34. Nitrate and nitrite fluxes were usually small and not found to be obvious functions of temperature. Since no orderly pattern for their magnitude could be determined, and since the direction of the fluxes roughly balanced to zero net release, the benthic flux of nitrogen oxides was not included in the model.

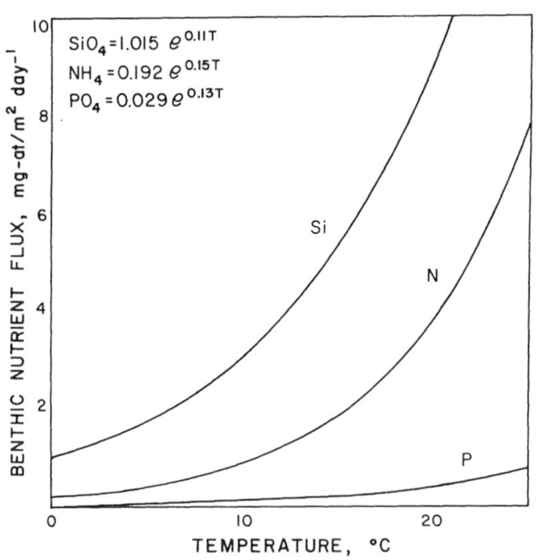

Fig. 34. Benthic regeneration of nutrients represented in the model as a function of temperature. Fluxes from three major bottom communities in the bay over an annual cycle were used to fit the empirical equations (Nixon et al., 1976 and in preparation) giving $mg-at/m^2\ day^{-1}$

The cycle of nutrients in the model is broken in the benthos. However, the cumulative balance of nutrients for the sediments is followed as a matter of interest. Sources to the sediment include the nutrients present in phytoplankton sinking to the bottom or consumed by benthic grazing and the unassimilated ration of zooplankton. These sources are balanced by the positive forced fluxes and cumulatively accounted for throughout the year. In this way the uncoupling of the nutrient cycle can be evaluated. The model equations for the nutrient fluxes into the water (FLUXAM, FLUXP, FLUXSI, μg-at l^{-1} day^{-1}) and the cumulative sediment balances (SEDN, SEDP, SEDSI, μg-at m^{-2}) are as follows:

$$\mathrm{FLUXAM} = \mathrm{FLXAM0} \cdot (e^{(\mathrm{FLXAMT} \cdot \mathrm{TEMP})})/\mathrm{LPERM2}$$

$$\mathrm{FLUXP} = \mathrm{FLUXP0} \cdot (e^{(\mathrm{FLUXPT} \cdot \mathrm{TEMP})})/\mathrm{LPERM2}$$

$$\mathrm{FLUXSI} = \mathrm{FLXSI0} \cdot (e^{(\mathrm{FLXSIT} \cdot \mathrm{TEMP})})/\mathrm{LPERM2}$$

$$\mathrm{SEDN} = \mathrm{SEDN} + (\mathrm{SINKN} + \mathrm{UAN} + \mathrm{RTNP} \cdot \mathrm{N{:}CP}-\mathrm{FLUXAM}) \cdot \mathrm{LPERM2}$$

$$\mathrm{SEDP} = \mathrm{SEDP} + (\mathrm{SINKP} + \mathrm{UAP} + \mathrm{RTNP} \cdot \mathrm{P{:}CP}-\mathrm{FLUXP}) \cdot \mathrm{LPERM2}$$

$$\mathrm{SEDSI} = \mathrm{SEDSI} + (\mathrm{SINKSI} + \mathrm{UASI} + \mathrm{RTNP} \cdot \mathrm{SI{:}CP}-\mathrm{FLUXSI}) \cdot \mathrm{LPERM2}$$

where: LPERM2 = DEPTH∗1000,
UAN, UAP, UASI = nutrients in unassimilated ingestion of zooplankton [Eq. (27)],
SINKN, SINKP, SINKSI = nutrients in sinking phytoplankton [see Sect. 4.3.5],
RTNP = benthic realized ration of phytoplankton.

6.4 Nutrients

Many aspects of the nutrient cycles in the model have been dealt with in detail in the discussions of individual compartments. Only the input of sewage, the process of nitrification, and the methods of accounting for nutrient changes remain to be described.

Urbanization, especially along the northern shores of Narragansett Bay, is extensive, and the input of nutrients in industrial and domestic sewage may be substantial. To preliminarily assess this role, effluent samples were collected from eleven major sources entering the bay on four occasions from July 1972 through August 1973. Samples were analyzed for nutrient concentrations and combined with flow estimates for the treatment plants yielding the approximate contributions to the bay system (Table 11; Nixon et al., in preparation). Sewage inputs in the model include only inorganic forms of silica, phosphorus, and nitrogen, although organic forms of nitrogen and phosphorus were also found in significant amounts. These organic fractions are not represented in the model, although some shortcomings of the simulated nutrient dynamics, as well as the appreciable levels of dissolved organic nutrients observed in bay water, suggest that this may be an important future addition to the chemical processes included in the model. At this

Table 11. Daily sewage contributions of inorganic nutrients to each element of the Narragansett Bay model based on analyses of effluent samples seasonally for 11 major sources and daily flow rate estimates (Nixon et al., in preparation)

Element	NH_4		NO_2+NO_3		PO_4		$Si(OH)_3$	
	$\frac{\mu g\text{-at}}{1}$	Total 10^9 µg-at	$\frac{\mu g\text{-at}}{1}$	Total 10^9 µg-at	$\frac{\mu g\text{-at}}{1}$	Total 10^9 µg-at	$\frac{\mu g\text{-at}}{1}$	Total 10^9 µg-at
1	0.698	90.7	0.210	27.26	0.093	12.14	0.0149	1.939
2	0.019	5.62	0.003	0.86	0.003	0.74	0.0005	0.156
3	0.015	1.70	0.010	1.13	0.004	0.45	0.0001	0.007
4	0.008	3.91	0.005	2.51	0.001	0.34	0.0002	0.081
5	—	—	—	—	—	—	—	—
6	0.030	6.68	0.006	1.35	0.004	0.85	0.0009	0.205
7	—	—	—	—	—	—	—	—
8	0.063	34.77	0.007	3.65	0.006	3.02	0.0008	0.474

point, there is far too little known about the nature, sources, or fates of dissolved organic compounds in sea water to attempt to formalize their role in the system.

Although uptake by phytoplankton of ammonia, as well as other nutrients, is diurnally periodic (Eppley, Rogers, and McCarthy, 1971), the generalization of average daily rates was used in the model. Since turnover may be quite rapid, approaching several times a day for ammonia in the bay (Vargo, 1976), an attempt was made to evaluate the simultaneous interaction of various rates. The total balance of all nutrient fluxes is combined in one calculation, so that one day's demand may be met by the same day's excretion and benthic flux. Considering also the predictor–corrector method of integration (Chap. 7), this calculation is probably satisfactory for turnover rates of less than or about once daily. However, during the midsummer, when small, rapidly growing flagellates may dominate the phytoplankton community, this scheme may be inadequate, and future modifications might incorporate a shorter or variable time-step.

For ammonia, sources include sewage input (SWGAM), copepod excretion of nitrogen not required to complement different C : N ratios of the food (EXN), ctenophore temperature-dependent excretion (CTEXN), and benthic flux (FLUXAM). These sources are balanced against total demand by the phytoplankton (DN). If the demand exceeds the available ammonia, the remainder is drawn from the nitrite–nitrate pool, as supplemented by any sewage additions. In this way, the model represents the NH_4 preference of phytoplankton, and the shifting of uptake constants for nitrate and ammonia (MacIsaac and Dugdale, 1972). Although it is technically possible for demand to exceed total supply, the predictor–corrector integration scheme compensates for this eventuality by reducing the growth rate accordingly.

The oxidation of ammonia to nitrite-nitrate is not included in the simultaneous supply–demand computation because the complicated mathematical expression required would have been unreasonably cumbersome to program. Instead, oxidation of ammonia occurs after the balance calculation. According to

Jaworski et al. (1972) nitrification may be described as a function of temperature by the equation:

$$NO_2 + NO_3 = NH_4 \cdot e^{(K_t \cdot \mathrm{time})}$$

where:

$$K_t = K_{20}\,\theta^{(\mathrm{Temp}-20)},$$
$$K_{20} = 0.068\ \mathrm{day}^{-1},$$
$$\theta = 1.188.$$

Phosphate and silicate balances are handled similarly, except that fewer processes are involved and the scheme is simplified. Sewage inputs and benthic fluxes, and excretion in the case of phosphate, are directly balanced against phytoplankton demand for the two nutrients. The model equations are presented below:

Net Balance of Sources:

```
      NH4 = NH4 + SWGAM + FLUXAM
            + EXN + CTEXN
   NO2NO3 = NO2NO3 + SWGON
      PO4 = PO4 + SWGP + FLUXP + EXCP
       SI = SI + SWGSIL + FLUXSI
```

Phytoplankton uptake:

```
           NH4 = NH4 – DN
         IF(NH4>0) GOTO 274
        NO2NO3 = NO2NO3 + NH4
           NH4 = 0.
274        PO4 = PO4 – DP
            SI = SI – DSI
```

Nitrification:

$$\mathrm{KNIT} = \mathrm{K20} \cdot (\mathrm{THETA}^{(\mathrm{TEMP-20})})$$
$$\mathrm{NITRAT} = e^{(-\mathrm{KNIT})}$$
$$\mathrm{DNH4} = \mathrm{NH4} \cdot (1\text{–}\mathrm{NITRAT})$$
$$\mathrm{NH4} = \mathrm{NH4} + \mathrm{DNH4}$$
$$\mathrm{NO2NO3} = \mathrm{NO2NO3} + \mathrm{DNH4}$$
$$\mathrm{NTOT} = \mathrm{NO2NO3} + \mathrm{NH4}$$

Simulation and Analysis

7. Mathematical Considerations and the Computer Program

Most of the chapters up to this point have been devoted to discussions of biological and ecological processes. In each case, a number of equations and coefficients have been derived for use in the simulation model. The form of the equations, however, is also important to the model and deserves some mention. Unfortunately, many of the biologists and ecologists whose work makes it possible to conduct mechanistic numerical simulations often turn away from examining or using such models because they are uncomfortable with the mathematical formalism and computer programming that are involved. In the hope of making the transition from laboratory and field data to coupled differential equations clearer, the following section reviews and summarizes many of the mathematical considerations that were involved in arriving at the form for the equations of the Narragansett Bay model.

Since a list of equations is far from being a simulation model, we have also described the development of sample sections of the computer program for the model. The intellectual effort and rigor of this aspect of modeling are often not appreciated by nonprogrammers, though the design of complex algorithms is a major part of simulation. A number of the more intricate sections of the Narragansett Bay program may also be of interest to experienced programmers because many of the same general kinds of programming problems appear in simulating other types of ecosystems.

7.1 Instantaneous Rates, Finite-Interval Rates, and Integration

The mathematical process of integration evaluates the result of a change in some dependent variable with respect to a change in an independent variable. The rate of change at any instant is represented by the derivative, dy/dx. After a specified change in the independent variable, e.g., time, the value of the dependent variable is defined by the initial value plus the integral of the derivative for that time interval. Thus, given a growth rate of phytoplankton (K, per day) and an initial population

(P_0, mgC l^{-1}), the population resulting after any number of days (P_t) is evaluated by integrating the change over time.

$$P_t = P_0 + \int_0^t \left[\frac{dP}{dt}\right] dt\,, \tag{40}$$

where: $\frac{dP}{dt} = K \cdot P$, the derivative of P with respect to t.

The form of the derivative determines what method can be used to solve the integration. By calculus some derivatives are known to have an exact solution, a single expression which calculates the integral with no numerical error. Many equations do not have this convenient property. Ecological processes, however, frequently are of a special class of growth equations for which an exact integral may be written. For example, in phytoplankton growth studies, the rate coefficient, K for the above derivative, is often defined by the following equation:

$$K = \frac{(\log_n P_t - \log_n P_0)}{t - t_0}. \tag{41}$$

K is frequently called the "rate constant", with the assumption that growth has been uniform during the time interval, $t - t_0$. The equation is more correctly thought of as a satisfactory approximation when K is a slowly varying quantity relative to the time interval. But if the assumption of constancy is met, K defines an *instantaneous rate* of change, and the exact, explicit integral may be calculated.

The base of the log [n in Eq. (41)] determines the biological meaning of K, i.e., base 2 gives K in divisions per day, or doublings per day. Notice that the inverse of $K_{\text{base}\,2}$ (div/day) is the doubling time, or generation time. For convenience in computer programming, the natural log (base e, ln) is often used. Rewriting Eq. (41) results in a familiar expression with widespread application in many areas of ecology.

$$K = \frac{(\ln P_t - \ln P_0)}{t - t_0}$$
$$\ln P_t = \ln P_0 + K(t - t_0)$$
$$P_t = P_0\, e^{K(t - t_0)}. \tag{42}$$

Note that the exponent of the base must always be unitless, so the units of the specific growth constant (base e) are time^{-1}, or "per day". Similarly, for metabolic processes:

$$\text{Resp} = \text{Resp}_0\, e^{K \cdot \text{temp}} \qquad (K\text{, per degree}), \tag{43}$$

or the physical extinction of light with depth:

$$I = I_0\, e^{-K \cdot z} \qquad (K\text{, per m—the extinction coefficient}). \tag{44}$$

The mathematical importance of using instantaneous rates to express changes is that the integrals are precisely determined by the Eq. (42) as long as the rate coefficient does not change during the time interval. In his early computational models, Riley (1947a) employed this exact integral form over successive two-week time intervals, during which he assumed constant daily rates of herbivore assimilation (A), respiration (R), carnivorous predation (C), and natural mortality (D). Thus the herbivore biomass two weeks hence ($\Delta t = 14$ days) was calculated:

$$H_t = H_0\, e^{(A-R-C-D)\Delta t}\,.$$

While the exact integral evaluation is not unknown in recent ecological models (Lassiter and Hayne, 1971), it is not the usual method.

Some equations cannot be explicitly integrated, and their progress over time must be approximated by successive summations of the changes over small time increments. For this, an alternative form of the derivative is used, the differential equation:

$$\Delta P = K \cdot P \cdot \Delta t\,.$$

An estimate of increasing standing stock is given by:

$$\begin{aligned} \Delta P &= K \cdot P_{t-1} \cdot \Delta t \\ P_t &= P_{t-1} + \Delta P\,. \end{aligned} \tag{45}$$

When the time interval, Δt, is small with respect to the units of K, the solution may be approximated by Eq. (45) with t stepping up iteratively toward the time at which the solution is required. Many schemes of *numerical integration*, such as Euler (Rectangular, as above), Trapazoidal, Adams, or Runge-Kutta, have been devised, which assess the integrated change more accurately than Eq. (45). They all converge to the exact integral when it is known. An advantage of the iterative method is that the rate coefficient, K, may be varied with each time-step, if necessary. In all such schemes of numerical integration, the accuracy of the solution may be increased by decreasing the time-step, Δt. Of course, this increases the computation time on the computer, and a satisfactory balance must be decided upon.

Because of the long-term simulations anticipated for the Narragansett Bay model, a short time-step was undesirable, and a time base of one day was chosen as reasonable in terms of the computation time involved and the time-frame of the biological rates under consideration. During periods of rapid change, however, a one-day step is too great for accuracy, so a predictor–corrector integration scheme was adopted. In effect this reduces Δt to one-half day, although only in a mathematical sense, and average 24 h conditions and rates are still assumed.

In the predictor–corrector scheme, rates of change for all processes are determined twice for each day, once with the initial values of all state variables, and again with the values that result after the predicted changes occur. The two rate estimates are then averaged, and the corrected computations are made updating the variables to new levels. Theoretically, the average may be determined assuming that the change between the first and second estimate follows a logarithmic curve:

$$\text{Rate}_{\text{avg}} = (\text{Rate}_2 - \text{Rate}_1)/(\ln \text{Rate}_2 - \ln \text{Rate}_1). \tag{46}$$

However, the additional computation time required for this embellishment is not warranted by the usually small improvement, and a simple arithmetic average was used.

As mentioned earlier, the exact integral of exponential rate processes is specified [Eq. (42)] if the rate coefficients do not vary during the integration time interval. Many biological processes are best expressed as instantaneous rates, and the integrated estimate of the daily change may be evaluated properly. In essence, then, the combined integration scheme assumes that average rates are valid for each 24 h period, but the exact integral of that rate is determined with no integration error. For a one day time-step, the improvement over a simple finite difference method is significant.

Example: if $K=0.693$ per day (1 div/day), and $\Delta t=1$ day,

Exact Integral:	Finite Difference:
$P=P_0 e^{K\cdot\Delta t}$	$P=P_0+K\cdot P_0\cdot\Delta t$
$P=2.000\,P_0$	$P=P_0+0.693\cdot P_0\cdot 1=1.693\,P_0$

The integration scheme used in the model is outlined below, for a simple case of phytoplankton growth (K) minus zooplankton grazing (g).

1. Determine rate coefficients as functions of forcing variables and state variables at time t.	$K=f(\text{Lt}_t, \text{Temp}_t, \text{Nutr}_t)$ $g=f(\text{Zoopl}_t, \text{Temp}_t, P_t)$
2. Predict changes due to these rates assuming constant conditions, and estimate new standing stocks (^).	$\hat{P}_{t=1}=P_t e^{(K-g)\Delta t}$ $\text{Nutr}_t \rightarrow \widehat{\text{Nutr}}_{t+1}$ $\text{Zoopl}_t \rightarrow \widehat{\text{Zoopl}}_{t+1}$
3. Determine rate coefficients as functions of forcing variables at time t but with estimated state variables for $t+1$.	$K^1=f(\text{Lt}_t, \text{Temp}_t, \widehat{\text{Nutr}}_{t+1})$ $g^1=f(\widehat{\text{Zoopl}}_{t+1}, \text{Temp}_t, \hat{P}_{t+1})$
4. Average the rates to give the corrected estimates.	$\bar{K}-(K+K^1)/2$ [or Eq. (46)] $\bar{g}=(g+g^1)/2$
5. Predict changes due to the corrected rates, using original state variables.	$P_{t+1}=P_t e^{(\bar{K}-\bar{g})\Delta t}$ $\text{Nutr}_t \rightarrow \text{Nutr}_{t+1}$ $\text{Zoopl}_t \rightarrow \text{Zoopl}_{t+1}$

An additional advantage of the exponential evaluation of the daily integrals concerns the problem of overshooting certain limits due to finite time-steps. For example, models frequently require checks that a predicted rate of consumption does not exceed the available supply, thus driving the source negative. Clearly this is a mathematical artifact. If the rate is evaluated exponentially, however, the source will never be depleted. Consider the case where temperature and initial food availability as detected by each individual copepod suggest a filtering of 100 ml day^{-1} $\mathrm{copepod}^{-1}$. The combined pressure of 10 copepods per liter will result in the exhaustion of available food, and more than 10 per liter will drive the finite difference calculation negative for the one day time-step. In the exponential treatment, numerical stability is insured, and the result is ecologically reasonable. In effect, a grazing threshold is implied, due simply to the increasing scarcity of the food. Let $g=0.1\ \mathrm{day}^{-1}$, for one copepod. With 20 copepods, the instantaneous rate will be $g=2.0\ \mathrm{day}^{-1}$. Considering the grazing effect alone, for $\Delta t=1$ day:

$$P=P_0\,\mathrm{e}^{-g}=P_0\,\mathrm{e}^{-2.0}=0.135\,P_0\,. \tag{47}$$

Thus, even a clearing rate of two times the volume daily leaves 13.5% of the phytoplankton remaining.

When two or more instantaneous rates are acting together, their simultaneous effects may be separated. Let the growth rate $K=0.693$ per day, and the grazing rate $g=0.600$ per day. Then, the combined effect is:

$$\begin{aligned} P&=P_0\,\mathrm{e}^{(K-g)\Delta t}\\ P&=P_0\,\mathrm{e}^{0.093}=1.097\,P_0\,. \end{aligned} \tag{48}$$

In this example the rates almost balance and the net change over one day is small (1.097). However, both growth and grazing are taking place and proper evaluation of nutrient uptake (resulting from growth) and zooplankton production (due to grazing) requires the separation of the effects. The appropriate equations may be shown from calculus to be:

$$\begin{aligned} \text{growth} &= K\cdot P_0(\mathrm{e}^{(K-g)\Delta t}-1)/(K-g)\\ \text{grazing} &= -gP_0(\mathrm{e}^{(K-g)\Delta t}-1)/(K-g)\,. \end{aligned} \tag{49}$$

The sign of the grazing term is negative by definition. Continuing the above example, for $\Delta t=1$ day:

$$\begin{aligned} \text{growth} &= 0.693\,P_0(\mathrm{e}^{0.093}-1)/0.093\\ &= 0.723\,P_0 \end{aligned}$$

and

$$\begin{aligned} \text{grazing} &= -0.600\,P_0(\mathrm{e}^{0.093}-1)/0.093\\ &= -0.626\,P_0\,. \end{aligned}$$

These two fluxes result in the same net change shown earlier in Eq. (48):

$$\begin{aligned} P&=P_0+0.723\,P_0-0.626\,P_0\\ P&=1.097\,P_0\,. \end{aligned}$$

Equations such as (49) are essential to the precise use of the exact integration method in ecological models. But they are quite complex compared to the component parts of a conventional differential equation. As mentioned earlier, most models to date have chosen to use numerical integration methods, tolerating their inherent computational error but retaining a more straightforward representation.

7.2 Developing the Computer Program

7.2.1 General Comments

Once the conceptual model is expressed in mathematical form, it must be incorporated into a rigorous computational framework or algorithm. In some cases, the development of this algorithm may require further simplifications or modifications in the model. The complexity of the program at any point is not always directly related to the complexity of the equation involved. Some of the most difficult and involved digital logic may be required to implement the intent of relationships which can be described with relatively simple mathematics. This section begins with a very basic introduction to the use of computers in simulation, and then provides samples from the Narragansett Bay program to illustrate the process of developing actual computer coding for simulation, as well as to describe our particular program in some detail. While this level of treatment can easily become tedious, we have tried to select particular pieces of the program that deal with relatively common problems in ecosystem modeling, or that address particularly intricate problems. It should also be pointed out that programming is not as divorced from the development of the conceptual model as it may appear from its being discussed here in a separate chapter. In fact, it is often the rigor of making the precise translation from flow diagram to equation to computer coding that exposes gaps or weaknesses in our understanding of the real system.

A first consideration is simply the choice of a language in which to write the program. Numerous options are open, each having particular strengths and weaknesses. FORTRAN, the language used in the Narragansett Bay model, was designed for scientific applications and, in addition, is perhaps the most widely used higher-level language. Others, such as PL-1 or APL, are less common, but have specific strengths that may be well suited to given applications.

Special simulation languages also exist, such as DYNAMO or IBM's Continuous Systems Modeling Program (CSMP), which are designed to minimize the supportive programming required to run a model. These languages provide a standard framework for representing the model structure, make available certain useful functional units which are easily incorporated into a model, and greatly simplify numerical integration. However, their generality results in increased overhead (in time and core requirements in the computer) and, more importantly, in the sacrifice of a complete knowledge of the ultimate structure and function of the computation module. For many applications, the ease and speed of achieving a workable program via these simulation languages is clearly desirable. But for extremely complex models, or those with an unusual time-lag or interfacing demands, such "package" languages may be unsatisfactory. For example, marine ecological models often must interface hydrodynamic and biological components. The time-steps required for suitable accuracy for the two may be on the order of minutes versus hours or even a day, respectively. While this is not a straightforward interfacing problem in any language, high-level simulation packages may be intolerably restrictive. Even if the difficulties can be surmounted, the package components may be combined in an unconventional way, potentially resulting in subtle "bugs" which may be extremely difficult to recognize and perhaps impossible to correct.

7.2.2 Digital Computer Simulation

A large part of this book discusses the derivation of mathematical formulations based on ecological data. In some cases, the translation of these expressions into computer coding requires only straightforward conversion into the FORTRAN language. In fact, throughout the discussions we have often used the FORTRAN statements as summaries to various section (e.g., pp. 58, 89, 103 etc.). Even to those with little or no practical programming experience, these statements are readily interpreted.

For example, recall the corrected time–depth integral for photosynthesis [Eq. (15)]:

$$\mathrm{LTLIM} = \frac{0.85 \cdot \mathrm{e} \cdot f}{k \cdot z} \left(\mathrm{e}^{\frac{-\bar{I}}{I_{\mathrm{opt}}} \mathrm{e}^{-k \cdot z}} - \mathrm{e}^{\frac{\bar{I}}{I_{\mathrm{opt}}}} \right)$$

This equation could be translated directly into FORTRAN. However, it is quite cumbersome to write this entire equation as a single FORTRAN expression, and it is frequently convenient to separate the calculations into shorter sections. In doing this, it is also desirable to avoid repeating identical subsections, since each computation step takes computer time and costs money. In this example, the multiplication $k \cdot z$ appears twice, as does the division $\bar{I}/I_{\mathrm{opt}}$. Thus it saves computer time to separate these calculations, and to store the results as intermediate answers by assigning them names. The names are arbitrary, and may be chosen to remind us of their meaning:

```
KZ = K*DEPTH
TERM1 = IBAR/IOPT.
```

In these FORTRAN statements, the two new variables, named KZ and TERM1, represent the answers that depend on the values of the other variables, which must already be specified. The * and / denote multiplication and division to the computer. Next we may define a third intermediate value, using the previous two:

```
TERM2 = TERM1*EXP(-KZ).
```

Here EXP is a special function recognized by the computer which specifies that exponentiation is to occur. Finally, the completed equation can be written:

```
LTLIM = 0.85*2.72*PHOTPD*(EXP(-TERM2)-EXP(-TERM1))/KZ.
```

The correct evaluation of this statement is insured because the part within the parentheses is completed before the last multiplication and division are performed.

A total simulation model includes a great deal more than the formulation statements. The working computer program must incorporate statements which specify coefficients, initialize the values of the state variables, control the flow of the program when it is not simply one after the other, and output the results in a useful form. While it is fundamentally very simple, it is not always initially obvious to those with little experience exactly how the computer may be used to simulate the behavior of a model system.

As an example, suppose we want to simulate the weight loss of an unfed organism due to its respiration under a slowly changing temperature regime, in this case an increase of 1° C daily. Our conceptual model assumes that respiratory losses are an exponential function of temperature, based on data which suggests a 10% loss daily at 0° C and a Q_{10} of 2.0. We choose to express respiration as a fraction of the body weight (lost per day) rather than as an absolute value. A simple simulation model in FORTRAN may be written using easily recognizable code names for the variables.

	FORTRAN Coding	Explanation
		Initializing values:
	RESP0=0.10	Physiological data, 10% loss of body weight per day at 0° C
	RESPT=0.0693	(ln 2)/10 to increase RESP0 with $Q_{10}=2$
	TIME=0.0	Arbitrary initial time
	DT=1.0	Values will be calculated daily
	ENDRUN=30.0	Simulation will be run for 30 days
	TEMP=0.0	Initial temperature = 0° C
	WEIGHT=100.0	Initial weight = 100
		Repeated iteration steps:
10	TIME=TIME+DT	Increment the time by DT
	TEMP=TEMP+1.0*DT	Calculate the temperature for this step (increases 1° C/day)
	RESP=RESP0*EXP(RESPT*TEMP)	Calculate instantaneous rate of respiration for this temperature
20	WEIGHT=WEIGHT*EXP(−RESP*DT)	Calculate the new weight
	PRINT 25, TIME,TEMP,RESP,WEIGHT	Print the results according to the FORMAT specified (25)
25	FORMAT (1X, 4F6.2)	Specifies the layout of printed results desired
30	IF(TIME.LT.ENDRUN) GO TO 10 STOP END	Check to see if 30 days have been run. If not (time less than 30), the calculations from (10) on will be run again

The execution of this program causes the computer to carry out each of these statements, in order, from the first to the last. However, upon reaching the statement numbered 30, the flow of the program depends on whether the end of the simulation has been reached. As long as TIME is less than the value of ENDRUN (.LT.30), the program branches back up to statement 10, repeating the calculations and printing the results for the next time-step. Should we wish to keep the temperature constant during the run, we could replace the constant 1.0 with zero in the temperature equation, or we could change the conditional branch to GO TO 20 instead of 10. When TIME finally equals ENDRUN, the branch-back does not occur, and the program stops executing.

Note that the equals sign in FORTRAN represents "is replaced by" rather than arithmetic equality. Thus, statement 10 replaces the present time with a new value, increasing by the assigned value of DT. Other values in the program must agree with the units implied by DT. The respiration coefficient RESP0 must be a daily rate, and ENDRUN must be in days. With internally consistent units, the value of DT may be varied, effecting the accuracy of the solution.

Statement 20 iteratively updates the weight, accomplishing the integration of the rate of loss through time. We are able to use this exact exponential integration due to the nature of this rate process (see pp. 107—110). More conventionally, statement 20 would be replaced by:

```
20  DW     = -WEIGHT*RESP
    WEIGHT = WEIGHT + DW*DT.
```

In this case, the answer would be more sensitive to the choice of DT.

The previous sample program was given to provide those with little computer experience some insight into simulation programming. Most useful simulations, of course, are considerably more complex. In many cases it is necessary, or desirable, to write the program in a way that may be quite complicated. Such coding complexities may be designed to use the computer more efficiently or to make the input and output options more convenient or meaningful. They may also be essential for the precise representation of the intended formulation. In any case, they ultimately determine the characteristics and the usefulness of a simulation model. For those with a knowledge of FORTRAN programming, a more technical discussion of the Narragansett Bay model follows. A complete listing of the computer program, along with samples of the data required for input as well as plotted output, is available from the authors.

7.2.3 The Narragansett Bay Computer Program

The Narragansett Bay Model is written in FORTRAN IV, for a G or H-level compiler. With small exceptions, the statements are within the ANS FORTRAN subset which is acceptable on even non-IBM FORTRAN compilers. With the entire model input from cards, a one-year standard simulation requires about 4 min and 256000 bytes of core for H-level compilation and execution on an IBM System

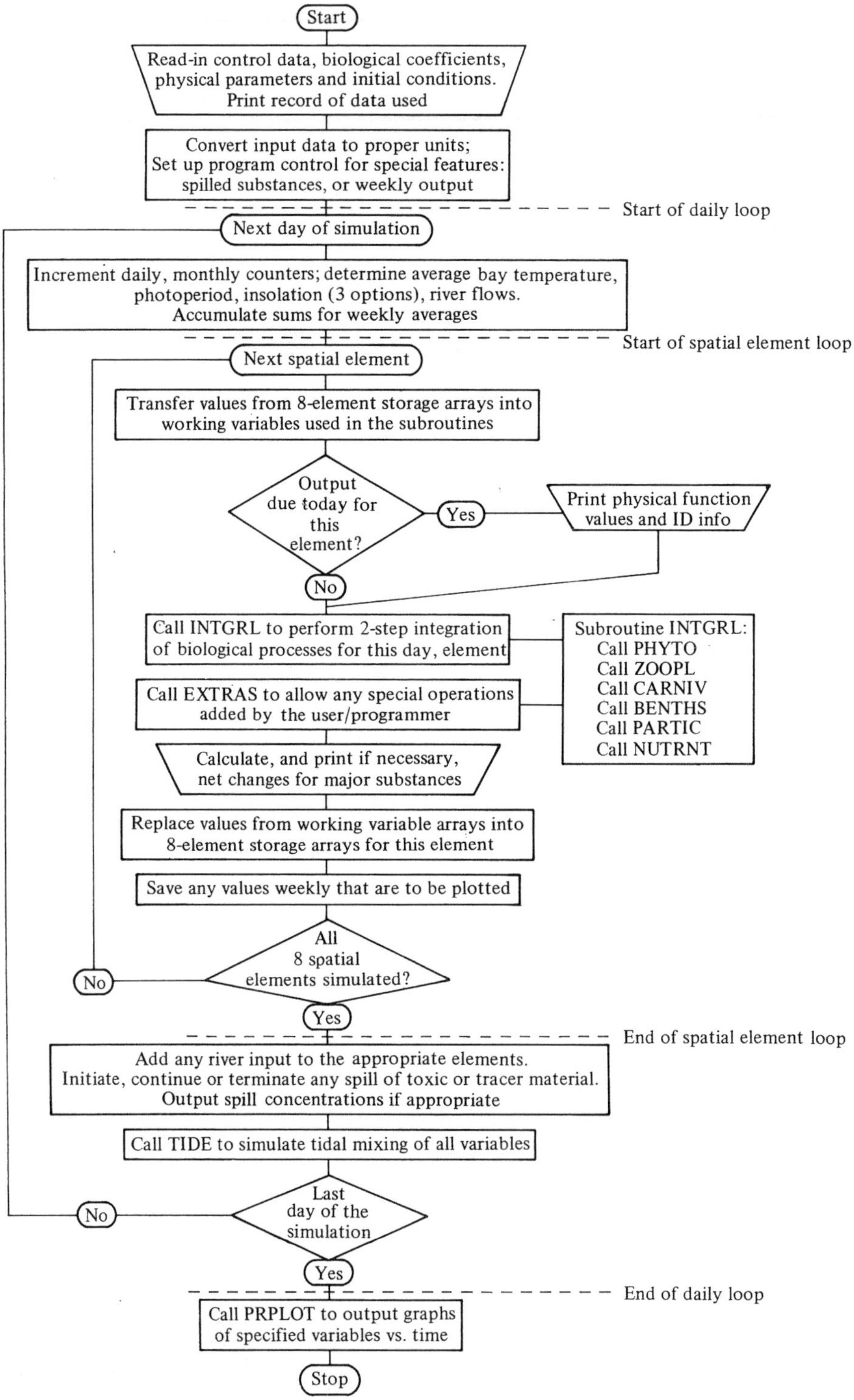

Fig. 35. Computer flow chart for the main program of the Narragansett Bay Model

370/Model 155. By preparing a precompiled module, the processing requirements are reduced to under 2 min and 192 K.

The program was designed in modular form with independent program units for conceptual clarity. The main program controls the input and output of data and results, determines the values of the forcing functions, assures proper sequencing in the iterations of time and space, and initiates the calling of the subroutines (Fig. 35). The conceptual relationships among the ecological compartments were presented in Chapter 2 (Fig. 8), and each compartment is represented in the program by a separate subroutine. The predictor–corrector integration is controlled by subroutine INTGRL, which, in turn, calls the ecological subroutines up to four times each day.

The remainder of this chapter consists of four examples taken from the Narragansett Bay Model that demonstrate some of the considerations involved in creating a precise yet reasonably efficient simulation of the conceptual formulations. The examples are arranged by increasing complexity.

Example 1. The Moving, Weighted Average for I_{opt} Acclimation

The theoretical and experimental bases for the assumption of I_{opt} acclimation were discussed in Chapter 4. The selected formulation designates the optimal light level for photosynthesis (I_{opt}) as a weighted average over the previous three days of the insolation at the one meter depth. The equation is simply:

$$I_{opt} = 0.7 \cdot I_1 + 0.2 \cdot I_2 + 0.1 \cdot I_3 . \tag{16}$$

To include this formulation accurately in the model requires some additions. First, as the model steps ahead in time, the correct insolation values must be retained and advanced each day so that the moving average remains properly weighted. Each day, after the calculation of I_{opt}, today's insolation value is stored, the value from yesterday and the day before are moved, and the value from three days ago is lost. Thus:

```
IOPT = 0.7*I1+0.2*I2+0.1*I3
I3 = I2
I2 = I1
I1 = SUBRAD*EXP(-K)          (today's insolation at 1 m).
```

Each iteration, the value of I3 is replaced by the value of I2, the previous value of I3 being destroyed. Note also that it is essential to arrange the statements in this reverse order (I3, I2, then I1).

Next, it is clear that these statements are impossible to evaluate on the first day of the simulation, since I1, I2, and I3 are as yet unspecified. Thus, special statements must be added to initialize these values. Since no single arbitrary value could be appropriate for all possible simulations, we chose to start the moving average with three identical values:

```
        IF(DAY.GT.1) GO TO 30
        I3 = SUBRAD*EXP(-K)
        I2 = I3
        I1 = I2
     30 IOPT = 0.7*I1+0.2*I2+0.1*I3
              ⋮ as before
```

After properly initializing the three terms on the first day, the program skips directly to statement 30 for all subsequent days.

Finally, the calculations are to be done for the eight regions of the bay, each with different physical characteristics and therefore potentially different plankton biomasses. Although insolation is the same all around the model bay, unequal phytoplankton populations result in varying extinction coefficients, and the three components of the moving average must be stored independently for the eight spatial elements. The DIMENSION statement in FORTRAN specifies that subscripted variable locations are to be reserved in the computer memory. The variable "ELT" becomes the subscript as the eight elements are simulated in turn:

```
        DIMENSION I1(8), I2(8), I3(8)
            ⋮
        IF(DAY.GT.1) GO TO 30
        I3(ELT) = SUBRAD*EXP(-K)
        I2(ELT) = I3(ELT)
        I1(ELT) = I2(ELT)
     30 IOPT = 0.7*I1(ELT)+0.2*I2(ELT)+0.1*I3(ELT)
        I3(ELT) = I2(ELT)
        I2(ELT) = I1(ELT)
        I1(ELT) = SUBRAD*EXP(-K)
```

The correct values for K and I_{opt} are computed and used only within the subroutine PHYTO, and therefore do not need to be stored with subscripts.

Example 2. Use of the EQUIVALENCE Statement to Facilitate Array Manipulations

Since there are eight spatial regions of the model bay, each with six primary state variables, and with the zooplankton further divided into adults, 12 egg compartments and up to 200 juvenile compartments, the number of values required to specify the state of the system at each moment is substantial. Thus it was important to develop as efficient a program as possible to manipulate these data.

In most of the program, it is clearly essential to distinguish each state variable from the others. For example, phytoplankton biomass must never be confused with an ammonia concentration! But within the subroutine TIDE, all substances are mixed in the same proportions, and there is no need to identify any given variable beyond knowing its spatial location. Further, since state variables must repeatedly be passed to and from the subroutine TIDE, maintaining them as separate entities would be time consuming. The FORTRAN EQUIVALENCE statement lets two names refer to the same stored array, allowing the programmer to choose between convenience and efficiency at will.

It is most advantageous to group the variables in small arrays, with subscripts denoting the identifying spatial region. Thus:

```
DIMENSION P1(8), P2(8), ZA(8),...
```

reserves storage for the two phytoplankton species-groups for the eight elements, the zooplankton adult biomasses, and so on. Since the various biological subroutines are written without regard to any location within the bay, the simulation of the eight elements is accomplished by passing the appropriate values of these small arrays to the subroutines. The state variables are maintained separately for easy identification.

However, these same variables may be grouped together in a single large array allowing all the variables to be efficiently passed to a subroutine, or any other uniformly applied operation:

```
DIMENSION X(8,230), P1(8), P2(8), ZA(8),...
EQUIVALENCE (P1(1), X(1,1)), (P2(1), X(1,2)), (ZA(1), X(1,3)),....
```

Note that the order of the subscripts of the large array is important. *X*(8,230) assures that all eight values of a single variable are contiguous. Only in this order is it possible to EQUIVALENCE the smaller arrays with the large one. Within the computer, the *X* array is stored in the following order:

X	(1, 1),	(2, 1),	(3, 1),	(4, 1),	(5, 1),	(6, 1),	(7, 1),	(8, 1),	(1, 2),	(2, 2),	...
P1	(1),	(2),	(3),	(4),	(5),	(6),	(7),	(8),	P2(1),	(2),	....

Clearly, the alignment of the smaller 8-part arrays is consistent with the layout of the large *X* array.

In this way, all of the state variables are readily referred to, including even the egg and juvenile zooplankton. Identified by name, EGGS (8,12) and ZJUV (8,200) are specified. But even these 2-dimensional arrays fit neatly into the composite *X* array.

Example 3. Enhancing Execution Efficiency by Streamlining Program Branching

Many details of a computer program may be required to control the sequential flow from statement to statement during execution. To facilitate this process, FORTRAN provides a number of branch control statements, whereby the direction of flow depends upon the value of a key variable. The "Logical IF" statement in the first example is frequently used and one of the most powerful branch statements. At anytime during execution, the value of the test variable may change, and program flow will be altered accordingly. However, in some cases the branching alternatives depend on test variables that are set once by initial conditions for the entire run. In such cases, it is inefficient to evaluate a Logical IF test at every iteration. This is a situation where additional programming can reduce the execution time resulting in time and money savings if the program is to be widely used.

The scheme used in our model employs a FORTRAN statement called an "Assigned GO TO". This statement type is the fastest branching method in FORTRAN, and allows the branching path to be defined once, early in the program. The Assigned GO TO allows branching to occur to any statement number previously specified. Consider this example:

```
      ASSIGN 10 TO JBR
        :
      GO TO JBR, (5, 10, 15, 20)
  5   X = 1.
        :
 10   X = 2.
        :
 15   X = 3.
 20   CONTINUE
```

The variable "JBR" is assigned a value designating the *statement number* to which the program will branch when the following GO TO JBR is reached. Here, the branch is to statement 10 (resulting in $X = 2$), but it could have been 5, 10, 15, or 20 since they were specified in the GO TO list. Once the designation of JBR is assigned, the branch occurs rapidly, with no conditional or logical test.

In the PHYTO subroutine, 64 variations of nutrient kinetic calculations may be required. The simplest case is a single algal species-group, with all three nutrient ratios constant (C:N, C:P, C:Si). The other cases include variable ratios for any one, two, or all three nutrients for either or both species. The actual coding is too lengthy to be reproduced here, but a partial listing will demonstrate the principle. Consider a simplified case where a single phytoplankton species-group may have variable nutrient-to-carbon ratios for any combination of the three nutrients: nitrogen, phosphous, or silica. Let the variables LUXN, LUXP, and LUXSI be indicators of whether or not each nutrient is to be variable during a given simulation run. LUXN $= 0$ when the C : N ratio is constant, but LUXN > 0 whenever luxury uptake kinetics are being tested, i.e., the ratio may vary during the run, requiring additional calculations each iteration to insure conservation of mass and the proper overall ratio within the phytoplankton population.

Perhaps the most straightforward approach to this programming problem would use "Logical IF" statements to branch around unnecessary calculations in the case where a ratio was not variable:

```
310  IF(LUXN.EQ.0.) GO TO 320
      ⋮ Special calculations for N-ratios
320  IF(LUXP.EQ.0.) GO TO 330
      ⋮ Special calculations for P-ratio
330  IF(LUXSI.EQ.0.) GO TO 350
      ⋮ Special calculations for Si-ratio
350  CONTINUE
      ⋮
```

Notice that although the special calculations are avoided when unnecessary (i.e., when LUX = 0), the Logical IF statements 310, 320, and 330 must be evaluated every iteration. We reduce execution time by using the more rapid Assigned GO TO statement that requires no logical test, thus streamlining the branching process. However, we must write an additional section to set up the flow pattern. Increased overall efficiency can result from a tradeoff of simplicity with longer execution time vs. additional program complexity and core requirements with more rapid execution. Using the "Assigned GO TO", the program appears as follows:

FORTRAN Coding	Explanation
GO TO JBR, (301, 310, 320, 330, 350)	
301 CONTINUE	Branch here only once.

ASSIGN 350 TO JBR
IF(LUXSI.GT.0) ASSIGN 330 TO JBR
IF(LUXP. GT.0) ASSIGN 320 TO JBR
IF(LUXN.GT.0) ASSIGN 310 TO JBR
	JBR is the first branch point, set to branch to the first special calculation required

ASSIGN 350 TO JBRN
IF(LUXSI.GT.0) ASSIGN 330 TO JBRN
IF(LUXP. GT.0) ASSIGN 320 TO JBRN
	JBRN controls branching immediately after the special *N*-ratio calculation if necessary

ASSIGN 350 TO JBRP IF(LUXSI.GT.0) ASSIGN 330 TO JBRP	JBRP controls branching after *P*-ratio calculations
GO TO JBR, (310, 320, 330, 350)	
310 CONTINUE ⋮ Special N-ratio calculations GO TO JBRN, (320, 330, 350)	
320 CONTINUE ⋮ Special P-ratio calculations GO TO JBRP, (330, 350)	
330 CONTINUE ⋮ Special Si-ratio calculations	
350 CONTINUE ⋮ End of special calculations, ⋮ Continue rest of program	

While the savings in execution time associated with this programming example are small for one-year runs of our eight-element bay model, they could easily become significant for applications with numerous spatial grids and/or longer simulations. Further, while any single improvement is almost certain to be small with today's ultra high-speed hardware, conscientious attention to efficient programming methods, such as those discussed here and in the previous example, can accumulate significant savings in costly execution time.

Example 4. Juvenile Zooplankton Growth, Development and Mortality

In the discussion of the zooplankton formulations (Sect. 5.8), we mentioned the increased complexity of the computer programming required to express the conceptual model. First, to ensure specific development times, the necessity for maintaining separate compartments for each day's production of juveniles multiplies the required calculations many times. Second, the interaction of temperature and available food delays juvenile growth to maturity by an unpredictable amount. And finally, the relatively complicated Eq. (49) required to conserve mass during the integration of growth, adult cannibalism, and carnivorous feeding obscures the program coding. For these reasons, it is of particular interest to examine in some detail how these features were represented in the model. But before presenting sections of the FORTRAN coding, it is necessary to define some of the pertinent variables.

ZJ(200). Within the computer, the biomasses of juveniles in all stages of development are stored in an array called ZJ, with a subscript denoting the number of days until sexual maturity is reached. The maximum subscript is 200, but in most simulations much of this array is empty, with development varying between one and three months.

D. Significant changes in temperature alter the rate of juvenile zooplankton development. Although the juvenile compartments represent homogeneous carbon pools in the model, the formulation of zooplankton development was facilitated by an extensive analysis of the pattern of growth of an individual from egg to adult. This analysis revealed a simple correction for the effect of constantly changing temperature on the predicted development time. Little error was introduced when the development time was taken as the average of an estimate based on temperature at the time of hatching and an estimate based on the temperature at the projected time of maturity. Such a correction is possible in the model since temperature follows a predictable pattern [Eq. (1)]. Thus, if on a given day the development time, D, was 50 days [Eq. (35)], and 50 days hence $D=20$ based on the regular temperature function, then the estimated development time used for today's juveniles would be 35 days.

ID, IDMAX. Since subscripts for computer arrays must be integers, the corrected development, D, is truncated to form another variable, ID. Upon hatching from eggs, each new crop of juveniles enters the ZJ array with the subscript ID, i.e., at the location ZJ(ID). Growth toward adulthood is then accomplished by transferring the numerical biomasses sequentially through a number of storage locations equal to the days of development. Finally, to avoid making numerous calculations for the many empty members of the ZJ array, a new variable is used to store the value of the longest development time that presently exists, IDMAX. At any moment, all juveniles in the entire bay are less than IDMAX days from becoming adults; all subscripts greater than ZJ(IDMAX) are empty.

IDSAFE. Adult cannibalism of juveniles may be designated to affect none, all, or part of the developing juveniles according to the value of the variable SAFE (p. 73). For example, if SAFE = 0.2, juveniles are safe from cannibalism during the last 20% of their development, approximately upon reaching copepodite Stage V. At any moment, IDSAFE is used to define the exact subscript value where cannibalism ceases:

$$\text{IDSAFE} = \text{SAFE} * \text{IDMAX}.$$

Thus, for a 35-day development and SAFE = 0.2, adult predation pressure is zero for array members ZJ(1) through ZJ(7), and nonzero for members ZJ(8) to ZJ(IDMAX).

FCARN, RLF, FTOT. The predation pressures from both the carnivores and the adult zooplankton are expressed as instantaneous filtering rates, FCARN and RLF, respectively. In the exponential integration method, their combined effect is the simple sum, FTOT = RLF + FCARN. Further, when multiple predators are competing for a common food supply, the quantity of prey received by each should be evaluated using the expression for simultaneous rates [Eq. (49)].

RTNZJ, TOCARN. As the predation on the juvenile compartments is calculated, the losses are partitioned between the adult zooplankton (RTNZJ) and the other carnivores (TOCARN).

ADV, IADV. Even with the correction for temperature variation, the development time assumes optimum conditions. The complete algorithm for delaying juvenile development also involves the fractional food limitation term, XRTNJ (p. 89). To simulate delay due to food scarcity, the transfer from one ZJ location to the next is only allowed to occur when the accumulated sum of the XRTNJ factor exceeds whole numbers, indicating the equivalent of one full day's growth. The integer IADV is set to 0 or 1 to signify whether the array is to be advanced. For example, if low phytoplankton concentrations reduced growth to 60% of the maximum for six consecutive days, juveniles would be transferred through the developmental array a little faster than every other day. Thus, for ADV = ΣXRTNJ:

Day =	1	2	3	4	5	6
XRTNJ =	0.6	0.6	0.6	0.6	0.6	0.6
ADV =	0.6	1.2	1.8	2.4	3.0	3.6
Advance =	No	Yes	No	Yes	Yes	No
IADV =	0	1	0	1	1	0

GRO. The final realized ration consumed by the juveniles is calculated by the exponential integration of the simultaneous instantaneous rates [Eq. (49)]. When this total ration is apportioned throughout all members of the juvenile array, the resulting growth increment is called GRO. This term is not an instantaneous rate,

	FORTRAN Coding	Explanation
		Define variables for predation by carnivores only:
	FTOT=FCARN	Total filtration rate due to carnivores
	EXFTOT=EXP(−FTOT)	Store temporary value, $e^{-\mathrm{FTOT}}$
	TERM=0.	Avoidingdivisionbyzerowhen
	IF(FTOT.NE.0.)TERM =GRO*(1.-FTOT)/FTOT	there are no carnivores, define temporary value for simultaneous rate evaluation [see Eq. (49)]
	FTERM=FCARN*TERM	Intermediate term to evaluate carnivores' share of ration
	GTERM=GRO*EXFTOT	Final net fractional change for each ZJ member, taking account of any predation and growth
		Begin computation for all ZJ members:
	DO 100 I=1,IDMAX	
	II=I+IADV	II equals either I or I+1, depending on ADV sum
	IF(II-IDSAFE)96, 92, 94	Is this ZJ member safe from cannibalism?
		For the 1st ZJ member subject to cannibalism plus carnivorous predation:
92	FTOT=FCARN+RLF	
	EXFTOT=EXP(−FTOT)	
	TERM=0.	Redefine terms for combined predation rates
	IF(FTOT.NE.0.)TERM =GRO*(1.-EXFTOT)/FTOT	
	FTERM=FCARN*TERM	Intermediate term to evaluate carnivores' share of ration
	ZTERM=RLF*TERM	Intermediate term to evaluate zoopl. adults' share of ration
	GTERM=GRO*EXFTOT	Final net change in ZJ
94	RTNZJ=RTNZJ+ZJ(II)*ZTERM	Accumulate ration to adults
96	TOCARN=TOCARN+ZJ(II)*FTERM	Accumulate ration to carnivores
	ZJ(I)=ZJ(II)*GTERM	Account for net biomass change in each ZJ member, *and* advance to next array location if appropriate
100	ZJTOT=ZJTOT+ZJ(I)	Accumulate total juvenile biomass

but a fractional increase by which each ZJ member would grow without any predation:

$$ZJ(I)=ZJ(I)*(1.+GRO).$$

These definitions have been purposefully brief, since many of them review concepts presented earlier in Chapter 5. The exerpt on p. 124 from subroutine ZOOPL of the bay model demonstrates the use of these variables in that portion of the program that updates the juvenile array ZJ, accounting simultaneously for carnivorous predation, adult cannibalism, and juvenile growth.

The critical point in accomplishing the proper choice of predation pressure is the branch control statement. For indices of ZJ below IDSAFE, the assignments of the initial section, appropriate to carnivorous predation alone, remain in force by branching to statement 96. When the first cannibalized ZJ member is reached, the branch to 92 redefines the intermediate computation values, and begins the accumulation of the adults' cannibalized ration. Thereafter, branching to 94 continues this accumulation.

In the complete program, this algorithm is a little more complicated. The ration accumulators (RTNZJ and TOCARN) must be initialized taking into account any predation on eggs, and the ZJ(1) member—which is maturing to adult status only if IADV=0. However, for the most part this section is a good sample of the sort of coding required to handle the multiple juvenile compartments. Similar equations are used for egg incubations, with the simplification that there is no delay by food limitation to contend with, and cannibalism is assumed always to occur.

8. The Tidal Mixing Model

Two simulations were run to evaluate some of the properties of the circulation scheme and mixing model. Both analyses involved following the concentration over time of a hypothetical tracer that was introduced into the bay. In the first case, a single release into the middle West Passage on 15 January was simulated. This resulted in a uniform concentration of 10.0 units/l throughout element 4. The second example simulated a slow but prolonged input of 0.6 unit in each liter of the

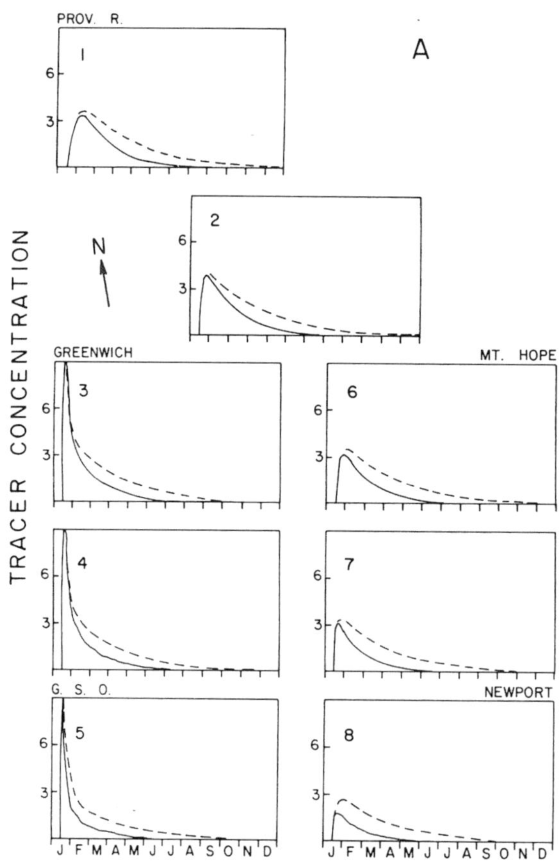

Fig. 36A

Providence River (element 1) for 100 days. This input reached a maximum concentration of 10.0 in element 1 on the final day of input. The time course of mixing and the decay for these situations is presented in Figures 36 and 37.

In the case of the instantaneous release, the primary weakness of the fast-mixing model, TIDE, is evident. In a very few days after the release, concentrations of 30% of the original 10.0 are found throughout almost all of the bay. While the compressed graphical time scale and the artificiality of this dye study exaggerate the problem (4.6 million kg would have to be homogeneously added into element 4 to result in an initial concentration of 10 mg/l), it is certainly true that a model with this crude a spatial resolution shunts material too rapidly between remote parts of the bay. Under normal conditions, however, where more gradual gradients exist, such accelerated exchange is a less-dominant feature of the model circulation.

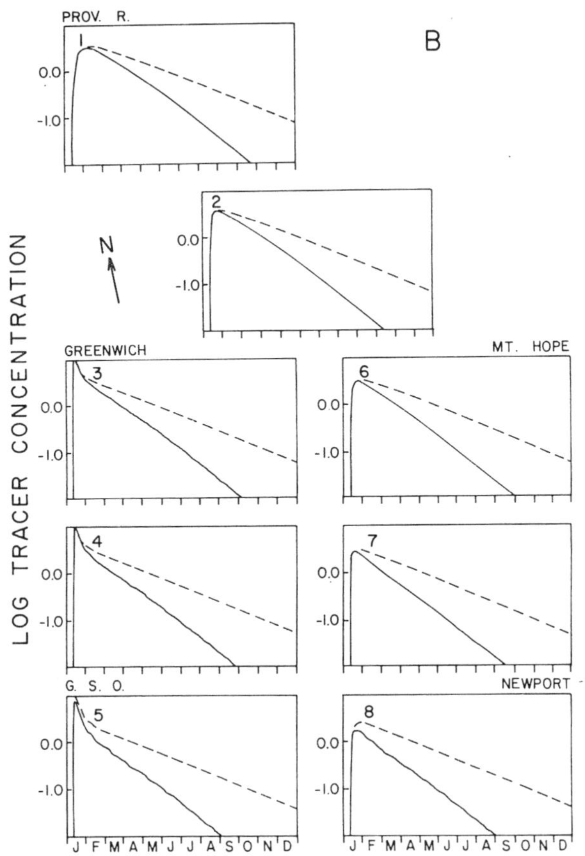

Fig. 36 A and B. Simulation of hypothetical tracer concentration in the eight spatial elements of Narragansett Bay using the rapid mixing model TIDE. A single instantaneous release raised the concentration of tracer to 10 units/l in element 4 (Fig. 1) on 15 January. Note the almost simultaneous appearance of tracer in all elements. *Solid lines:* simulated concentrations using high exchange estimates for the West and East Passages with Rhode Island Sound; *broken lines:* simulation using minimum estimates (Table 6). (A) Simulated concentrations through time. (B) Logarithm of the simulated concentration through time

After the initial dispersion of the tracer, the more fundamental flushing characteristics of the model elements emerge. This is particularly evident in the logarithmic plot of concentration for the instantaneous release study (Fig. 36B). The initial rapid dispersion phase is marked by the steep slope in the West Passage for two to three weeks following the release. After that, a relatively gradual and constant slope results, as the residual tracer is washed from the system.

These graphs also clearly demonstrate the significance of the boundary exchange factors. In all the figures the solid line represents the higher exchange

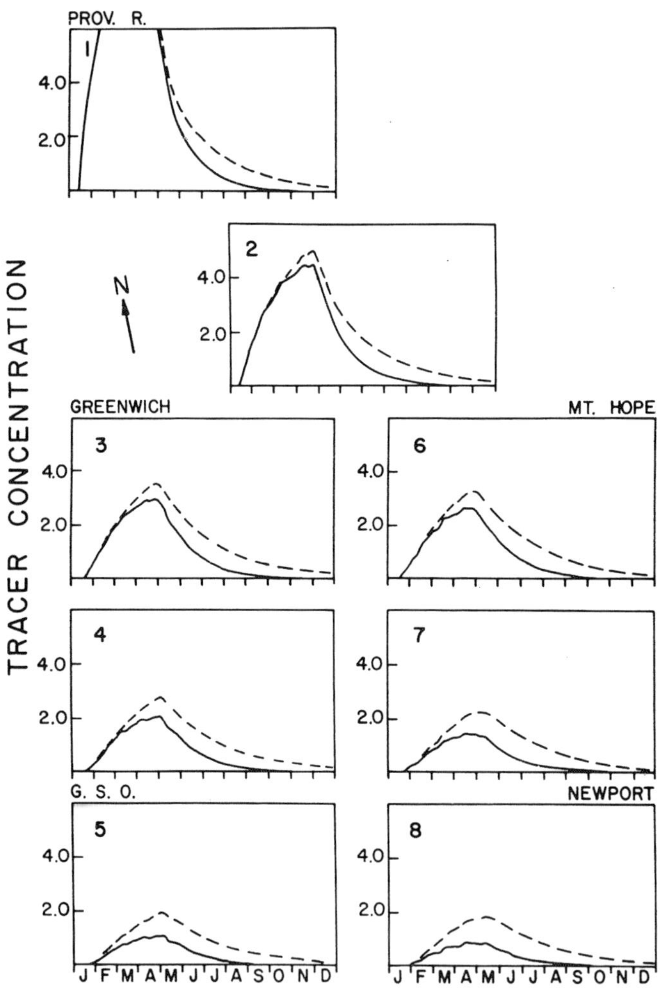

Fig. 37. Simulation of hypothetical tracer concentrations in the eight spatial elements of Narragansett Bay using the rapid mixing model TIDE. Beginning on January 15, the tracer was input at a daily rate of 0.6 units/l for 100 days in the Providence River (element 1). *Solid lines:* a simulation using high estimates of tidal exchange with Rhode Island Sound; *broken lines:* a simulation using minimum estimates of offshore exchange (Table 6)

estimates of 0.6 and 0.7 for the West and East Passage, respectively (see Table 6). The dashed curves are for WRIS = 0.8 and ERIS = 0.9, the minimum flushing estimates. In the ecological simulations to follow, intermediate values of 0.7 and 0.8 were used. The differences in flushing are significant for this range of factors. Using the slope of the log plots to calculate the instantaneous flushing rate [$m = \log_e C/C_o$)/time] indicates that this decay ranges between −0.0236 and −0.0124 day^{-1} for the high and low exchange factor estimates, so that the uncertainty is almost a factor of two for these very rough estimates. Stated another way, the dilution due to tidal flushing alone may be expected to reduce the concentration of a conservative property by 50% every 30 to 55 days. According to this model, as well as other estimates, the residence time in the bay is quite prolonged.

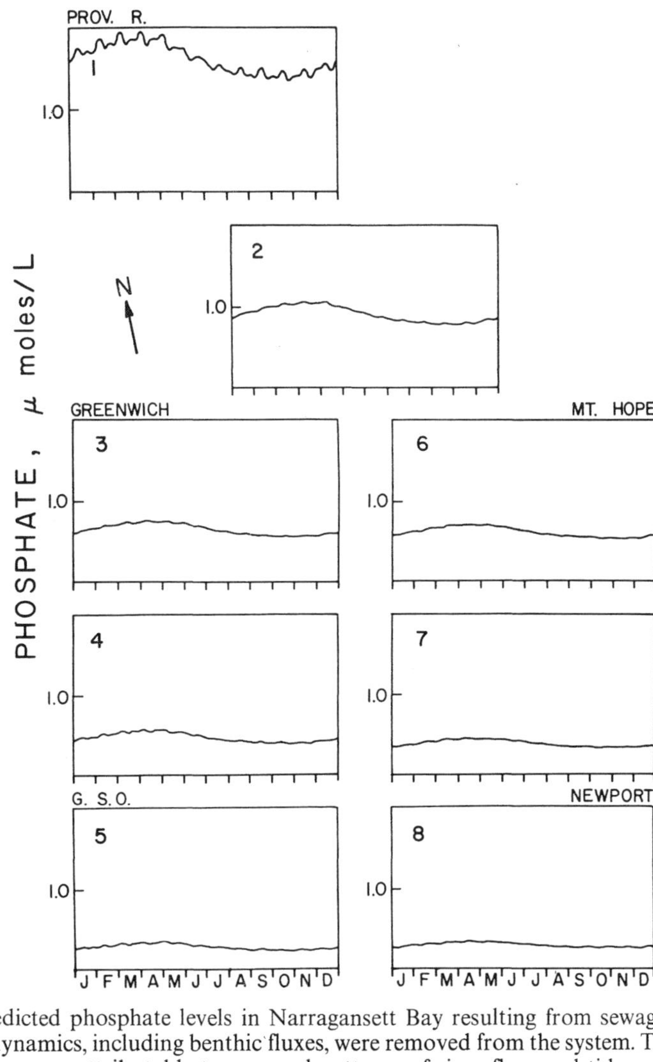

Fig. 38. Predicted phosphate levels in Narragansett Bay resulting from sewage inputs. All biological dynamics, including benthic fluxes, were removed from the system. The variations are attributable to seasonal patterns of river flow and tides

Table 12. Comparison of predicted (TIDE) and observed (Schulz, 1974) distributions of suspended hydrocarbons and particulates in Narragansett Bay over an annual cycle

Model element	Σ Hydrocarbons, μg/l		Σ Particulates, mg/l	
	Observed Schultz	Computed TIDE	Observed Schultz	Computed TIDE
1	50–100	Forced (50 and 100)	5–12	Forced (5 and 12)
2	28.7	20 –40	7.2	2 –4.8
3	18.6	11.5–23.1	8.2	1.2–2.8
4	7.1	6.6–13.1	4.9	0.7–1.6
5	2.8	2.1– 4.1	3.5	0.2–0.5
6	4.5	4.7– 9.5	3.0	0.5–1.1
7	3.7	2.5– 5.1	2.2	0.3–0.6
8	—	0.7– 1.4	—	0.1–0.2

The second tracer study, the 100-day continuous input into element 1 (Fig. 37), depicts similar features. The more moderate input appears gradually around the bay, approaches steady state and, when the source is stopped, begins to decay. The minor fluctuations are due to the daily variation in the tidal input forcing the model.

While it was expressly designed for use in the ecological model, TIDE is easily applied to other problems. For example, to assess the role of sewage inputs to the nutrient patterns in the bay, a simulation was run with all biological processes removed, so that only external nutrient sources and tidal mixing were included (Fig. 38). As expected, the patterns were uncomplicated, especially since no seasonality of nutrient inputs was assumed. The model was run for two years to allow the system to reach steady state. Two cyclic patterns are obvious, and both are due to physical factors. Short-period variations in the forced tide height are superimposed upon the yearly seasonal pattern of river flow. In addition, the run provides an indication of the approximate levels of nutrients attributable to sewage sources at steady state.

In another application, Schultz (1974) used the model to test the assumption that tidal mixing of suspended hydrocarbons introduced into the Providence River accounts for their observed distribution in Narragansett Bay. The close agreement between his observations and the model support this interpretation, while the disagreement with particulate matter distribution confirms that other major sources of particulates may be involved (Table 12). In addition, TIDE has been employed by Resource Economists at URI in a preliminary assessment of effluent loading patterns that may develop under various projections of economic and industrial growth for the state. While the gross spatial detail and the resulting accelerated appearance of small quantities of substances throughout the bay represent limitations, the extreme speed and simplicity of the subroutine make TIDE useful not only for the ecological application, but also for other tasks where first approximations of tidal mixing are required.

9. The Standard Run

9.1 Selection of Parameters

For each simulation with the model, over 50 physiological coefficients, about 10 baywide or seasonal patterns for forced compartments, and initial conditions throughout the bay for the six state variables must all be specified. These numerous variables and options preclude an exhaustive presentation of all the features of the model. To facilitate the discussion, one "standard" simulation run has been chosen with which to explain the basic input requirements and output results, and against which a number of modifications will be compared. The majority of coefficients and specifications have not been routinely altered in the simulations. For example, the biomasses and rates for the secondary biological compartments were not varied in most cases, nor were the flux rates, external nutrient sources, or physical properties of the bay. The selection and justification of these values have been presented in the respective formulation discussions.

The primary thrust of the Narragansett Bay model centered on the plankton system, and it is the coefficients and rates in these formulations that have received the most attention. In many cases the choice of a single value was difficult due either to wide disagreement in the literature, or to the knowledge that seasonal community changes probably make one average value inappropriate for the entire year. Despite

```
&ZOO IVLEVK= 7.0,RMXO=0.25 ,Q1ORMX=1.80,XRSPO=0.10 ,Q1ORSP=2.0 ,
     C$NZ= 5.0,  C$PZ= 85.0,  XASSIM= 0.8, C$DWZ= 0.35,  ZJGMXO= 0.03,
     JZMAX= 150,  SAFE= 0.2,  RJ$RA= 1.44,  &END

&PPL  C$N=  7.0, 0.0,  10.0, 5.0,  C$P= 85.0, 0.0,  85.0, 0.0,
     C$SI= 15.0, 5.0,  15.0, 5.0,  C$CHL1= 30.0,  C$CHL2= 30.0,
     KNP1=1.5,  KNP2=1.0,  KPP1=0.05,  KPP2=0.05,  KSIP1= 0.5,
     KSIP2= 0.05,  SIOP1=0.0,  SIOP2= 0.0,  TFORM= 1,  SNKRAT= 0.0,
     REQIAV= 40.0,  &END
```

Fig. 39. Values of physiological coefficients for zooplankton (*top*) and phytoplankton (*bottom*) in the "standard run". The choices are shown as the actual cards input to the computer using the FORTRAN NAMELIST option. Note the unambiguous identification of each parameter value and the conveniently flexible format provided by this data input option

these problems, estimates for the phytoplankton and zooplankton coefficients for the standard run were made as circumspectly as possible. These choices, in the form in which they are provided to the computer (Fig. 39), are presented below, with a brief description of their meaning and derivation.

9.2 Zooplankton

IVLEVK = 7.0 l/mgC—the exponent of the food limitation equation [Eq. (24)]. This term controls the availability of phytoplankton and particulate carbon food to adult and juvenile zooplankton. The value was chosen as representative for Narragansett Bay, although a wide range of values certainly exists (Fig. 28; Table 8).

RMX0 = 0.25 day^{-1}—the maximum adult daily ration at 0° C [Eq. (23)]. The value of 25% of the body carbon was chosen relative to the respiration rate below, and falls well within the range of reported values (Table 8).

Q10RMX = 1.8—the physiological Q_{10} for adult daily ration [Eq. (23)]. A value slightly less than 2.0 was chosen, resulting in a gradual decrease in metabolic efficiency with respect to respiration with warmer temperature. This assumes that increased behavioral activities would decrease the fraction of the adult ration available for growth.

XRSP0 = 0.1 day^{-1}, Q10RSP = 2.0—the terms of the temperature response for adult respiration [Eq. (28)] were chosen based on the observations of Conover (1956, Fig. 19B) on *Acartia*. A regression including cold-water observations for *A. clausi* and warm-water observations for *A. tonsa* suggested a 0° C respiratory rate of 2.64 $\mu l\, O_2\, mg^{-1}\, h^{-1}$ and a $Q_{10} = 1.99$ ($r^2 = 0.92$). Assuming 35% of the dry weight is carbon and 2 g O_2 per gC respired, the calculated metabolic rate is 0.13 mgC mgC^{-1} day^{-1}. The choices of 0.1 and 2.0 reflect these approximations of a seasonally acclimating population.

C$NZ = 5.0, C$PZ = 85.0 ratio by atoms[1]—estimates of the elemental composition of zooplankton are quite variable. These C:Nitrogen and C:Phosphorus atomic ratios were chosen based on a variety of observations (see Sverdrup et al., 1942; Ketchum, 1962; Conover and Corner, 1968; Mayzaud, 1973) and represent a slight protein enrichment over phytoplankton.

C$DWZ = 0.35—the carbon to dry-weight ratio is used in the model to convert the empirical initial conditions of dry weight into the model units of carbon. This value is representative for mixed crustacean zooplankton (see Curl, 1962).

XASSIM = 0.80—a high and constant assimilation efficiency was selected based on numerous observations using a variety of methods (see Chap. 5).

ZJGMX0 = 0.03 day^{-1}; JZMAX = 150 days—the selection of the maximum juvenile growth rate and development time are based jointly on considerations of egg and adult sizes (see Chap. 5). The temperature coefficients (Q_{10}) have been fixed at 0.1 and the 3% daily rate results in a growth from an egg weighing about 0.1 µg (for *A. tonsa*, Heinle, 1969a) to an adult weight of about 6 µg (Conover, 1956; Petipa, 1966; Heinle, 1969a) in a reasonable development interval.

---

[1] The symbol "$" is used in the program to denote conversion ratios since the colon is not accepted in variable names; e.g., C$N should be interpreted as C:N.

SAFE = 0.2—this value assumes that adults are effective predators of juveniles during most of the growth period. Only those juveniles that are within 20% of their development time of becoming adults are not cannibalized.

RJ$RA = 1.44—using allometric calculations for time-averaged juvenile versus adult weight, a factor expressing the increased respiration of the smaller juveniles may be calculated (see Chap. 5). Here, the rate for juveniles is simply 1.44 times the temperature-dependent adult respiration (mgC mgC^{-1} day^{-1}).

9.3 Phytoplankton

C$N = 7.0, ratio by atoms. A great variety of values has been reported for the C:N ratio in many phytoplankton species (see Vinogradov, 1953; Parsons et al., 1961; Strickland, 1966; Eppley et al., 1971; Burkholder and Doheny, 1972). An average of cited values for diatoms was 6.9 ($\sigma = 3.15$), with a somewhat higher value, perhaps as high as 10, for flagellates. The choice of any single constant value for nutrient composition is particularly inappropriate. Thus, two species-groups with unequal nutrient kinetics, as well as a provision for a variable-ratio, luxury uptake scheme may be included and will be discussed later. The 2 pairs of numbers for each nutrient ratio indicate which options are in effect.

C$P = 85.0, ratio by atoms. This value was selected from a similarly varied set of observations. An estimate for *S. costatum* was 54 (Parsons et al., 1961), a mixed estuarine population was 117 (see Strickland, 1966), and the average for nine estimates was 84.8.

C$SI = 15.0–5.0 (by atoms). The choice of a constant C:Si ratio is clearly inappropriate for a system dominated by diatoms during the colder months and by flagellates in the summer. For this reason, a seasonally varying ratio was modelled so as to track the temperature, reaching the maximum C:Si, e.g., 15.0, at 20° C and the minimum, 5.0, at 0° C. A similar result is obtained in the model by specifying a two-species system, one with a reduced silica requirement (see Chap. 4). The literature is in some disagreement on the appropriate silica content of diatoms. Some of this may be due to the apparent temperature effect resulting in increased frustule size in colder water. Some observations (Mitchell-Innes, 1973; Paasche, 1973a) suggest a minimal cell silica requirement which is insufficient to account for the observed drop in silicate concentration in Narragansett Bay during the spring bloom. However, recent analyses for natural bay populations support a more consistent range of values around 6–13 C:Si by atoms (E. Durbin, Phytoplankton Ecology Group, U.R.I., personal communication).

C$CHL1 = 30.0, C$CHL2 = 30.0, ratio by weight. The ratio carbon:chlorophyll is also very variable (see Steele, 1962). Parsons et al. (1961) reported a ratio of 26 for *S. costatum*, with values of 40 to 50 for another diatom and dinoflagellates, respectively. Ratios observed for natural samples are often substantially higher, reflecting the presence of other particulate carbon contributions. The value of 30 was chosen to represent an actively growing, small diatom population, although higher values might also be justified. Within the model, this value is used only to convert the initial conditions (mg Chl/l) into the program units of carbon, and in the self-shading calculation for the extinction coefficient (Sect. 4.3). Thus, the dynamics of the model are not sensitive to this choice.

Additional detail could be added to allow for analyses of changes in this ratio, as well as in the assimilation number (carbon growth/Chl). The present value is chosen to be representative of a *Skeletonema*-dominated diatom population (Parsons et al., 1961). Unequal ratios may be specified for two species-groups when this option is chosen.

KNP1 = 1.5, KNP2 = 1.0 μg-at N/l. The half-saturation constants for growth and nitrogen uptake have been widely studied (Sect. 4.2). Reported values usually range between 1 and 3, with large estuarine species generally characterized as being less efficient at lower concentrations.

KPP1 = 0.05, KPP2 = 0.05 μg-at P/l. The role of phosphorus in the bay is probably less significant than nitrogen or silica, making the choice of the half-saturation constants less critical. This value is based on observations of uptake and subsequent calculations for growth of various species ranging between 0.0008 and 0.07 (Hanton, 1969).

KSIP1 = 0.5, KSIP2 = 0.05 μg-at Si/l. Paasche (1973, 1973a) has reported half-saturation constants for the silicate kinetics of five diatoms. Values ranged from 0.08 to 3.7, and the value of 0.5 is weighted low to represent a *Skeletonema*-dominated population. The low value chosen for the second species-group is consistent with the characterization of a summer population requiring less silica for active growth.

SI0P1 = 0, SI0P2 = 0 μg-at Si/l. The presence of a lower threshold for silica uptake has been reported (Paasche, 1973a). While included in the program for completeness, these thresholds were not used routinely in the simulations.

TFORM = 1. For simulations involving one species-group, or when both groups respond equally to temperature, this control option is set to 1. TFORM = 2 results in a modified thermal response for the second species-group whereby its growth is slower than the Eppley (1972) curve in winter, and faster when the temperature exceeds the yearly baywide average of 11.5° C (Fig. 15A).

SNKRAT = 0. m/day. For most model runs, the sinking rate was zero with the assumption of complete vertical mixing. When a positive value is specified, an appropriate proportion of the first species-group is lost to the benthos each day.

REQIAV = 40.0 average ly/day. A lower limit to the light acclimation by the phytoplankton is an intuitively reasonable constraint. Riley (1967) observed that bloom inception in Long Island Sound seemed inhibited when the average insolation for the water column was below 40 ly/day. Our observations suggest that acclimation apparently maintains an optimum for photosynthesis (I_{opt}) equivalent to the light at about 1 m depth (Sect. 4.3). Accordingly, in the model the time-delayed acclimation of I_{opt} is constrained above the insolation at 1 m which results in the specified minimum average (REQIAV) for the water column. For example, let the extinction coefficient $k = 0.75\ \mathrm{m}^{-1}$, depth $z = 10$ m and REQIAV = 40.0 ly/day. From the equation of Riley (1967):

$$\bar{I} = I_0(1 - e^{-kz})/kz .$$

The surface insolation I_0 resulting in $\bar{I}=40$ may be calculated to be 300 ly/day, or, at 1 m, $I_1=142$. Thus, for these conditions I_{opt} would not be permitted to go below 142 (LOIOPT = 142., see Sect. 4.3).

9.4 General Description

The observed temporal and spatial patterns of distribution for plankton and nutrients calculated by the model were presented in Chapter 1. The standard simulation models the simplest phytoplankton option, a single species-group with constant nutrient kinetic properties throughout the year. Considering this gross simplification, the agreement with observed data from the baywide Systems Ecology Program is encouraging. The difficulties during the latter part of the year will be discussed shortly, but first the general consistency of the predicted relative gradients around the bay should be noted.

Significant differences in the concentrations along the north–south axis of the bay are well reproduced, suggesting that tidal circulation and flushing are primary factors in determining the spatial patterns in the bay. The Providence River (element 1), however, is an exception. In most cases the model is less satisfactory for this region. Phytoplankton are generally over-estimated, and nutrients under-estimated. Apparently some additional factors are critical in controlling ecological processes in this region. Such a conclusion is not unreasonable considering the substantial inputs of potentially toxic substances in the effluents flowing into the river.

The timing and magnitude of the winter–spring phytoplankton bloom are well represented in the model (Fig. 40A). The termination of the bloom is followed by a decline to a relatively stable plateau for a month or so, and the onset of a period of oscillations in the late summer. A slight decline occurs into November, and the beginning of a bloom appears evident early in December, especially in the upper elements.

The observations reflect a similar pattern for Narragansett Bay. The bloom in 1973 was delayed until late February, resulting in another counter-example to the "typical" winter–spring inception (Smayda, 1957; Pratt, 1959, 1965). The multiple oscillations following the spring plateau were substantially more extreme in nature than in the model, especially in the upper bay. The multiple peaks reflect the influence of small flagellate populations turning over rapidly in the warm waters (Nixon et al., in preparation; E. Durbin, personal communication). The bay phytoplankton demonstrated substantial depression as early as August, with low biomass throughout the fall. This pattern is only approximately matched in the model. The absence of a fall bloom has been observed before, and is particularly interesting since other factors would seem to suggest that active growth should be possible. In many model simulations, the summer oscillations continued into the fall, depicting small blooms as late as November. While this is later than the more typical fall peak, even such a delayed flowering has been observed in the bay. A striking feature of the seasonal pattern is the relative magnitude of the spring and summer blooms. Although the summer pulses are generally characterized as smaller (Pratt, 1959;

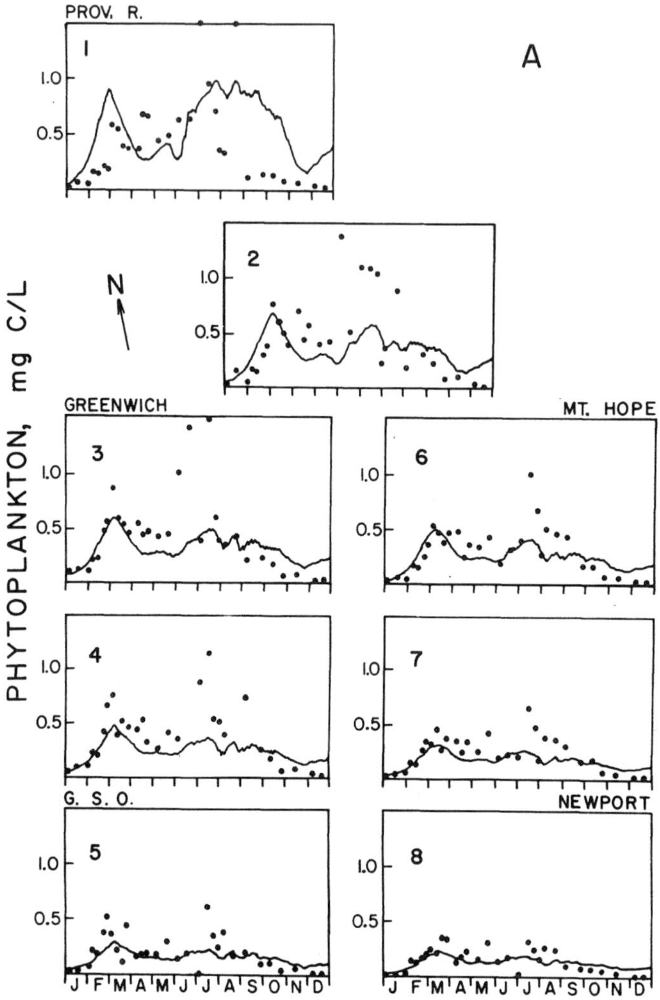

Fig. 40 A–E. Comparison of computed *(solid and broken lines)* and observed (●) values in the eight spatial elements of Narragansett Bay (Fig. 1) for the standard run and sample year Aug. 1972–Aug. 1973 (Figs. 4–6). (A) Phytoplankton. (B) Zooplankton adults plus juveniles over 50% mature *(solid line)*, and total zooplankton *(broken line)*. (C) Silicate. (D) Total dissolved inorganic nitrogen *(solid line)*, and ammonia *(broken line)*. (E) Phosphate

Smayda, 1973a), these observations suggest the reverse was the case this particular year. Since the data presented here as carbon more precisely reflect chlorophyll, a more reliable carbon estimate would consider variations in the C:Chl ratio between the winter and summer communities. In any case, the importance of these summer crops is substantial with respect to the earlier bloom.

Two components of the model zooplankton compartment are presented in Figure 40B. The dashed line portrays the total adult and juvenile biomass, while the solid line excludes younger juveniles less than 50% mature. Since nauplii and even

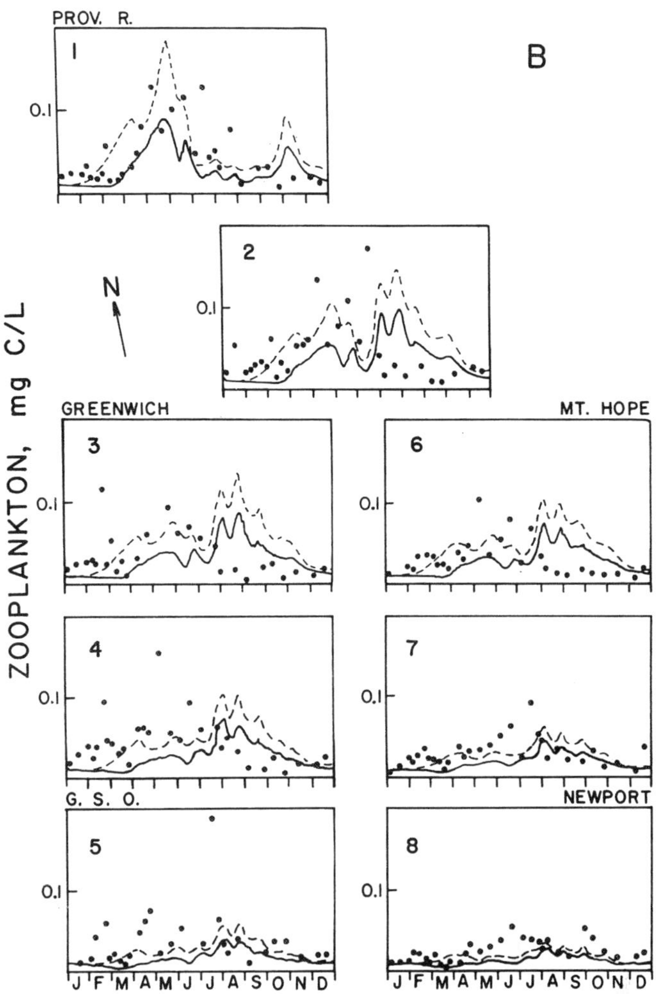

Fig. 40B

early copepodite stages were not well sampled with the No. 10 net used in the tows, the observed data may be expected to represent only a portion of the total natural population, probably intermediate to that represented by the two simulated lines. The brief February peak of observed zooplankton in the West Passage preceded, or in some cases coincided with, the maximum diatom flowering. This, combined with cold-water temperature, suggests that there was a release of meroplankton, perhaps barnacle larvae (Martin, 1965; Hulsizer, 1976), rather than a pulse of copepods. Even attributing this peak in the West Passage to meroplankton, which the model cannot be expected to produce, the levels of zooplankton remain quite low during the first two simulated months.

In all runs to date, the model has consistently indicated that the prebloom winter phytoplankton biomass is insufficient to maintain even the low observed zooplank-

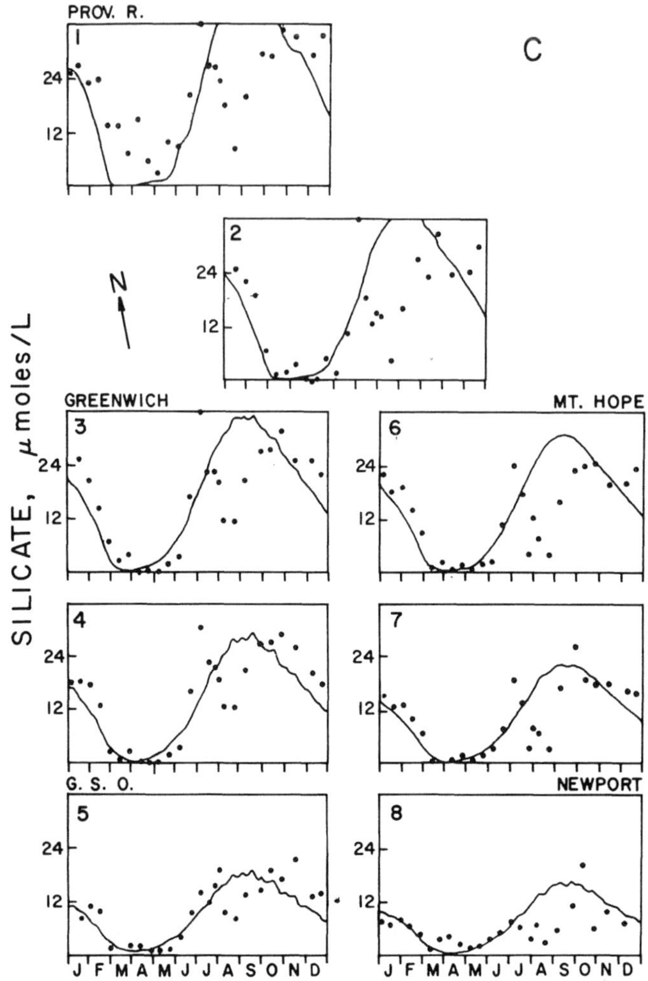

Fig. 40 C (legend see page 136)

ton stock. This suggests that alternative food sources are almost certainly being utilized during this time, a conclusion which is supported by observations of the general omnivorous nature of *Acartia* and other estuarine copepods (Petipa, 1966; Corner, 1972; Gerber and Marshall, 1974). Since the model was routinely run for one-year simulations beginning January 1, it was necessary to compensate for this nutritional deficiency so that the zooplankton in the model would maintain appropriate levels during the late winter. Such adjustments are not viewed as weaknesses of the model, since it is precisely indications such as these which make the model useful in adding to our understanding and directing future research.

To meet this deficiency, two compensations were adopted in the lower bay. Whenever zooplankton biomass in element 5 or 8 fell below 0.01 mgC/l, a value

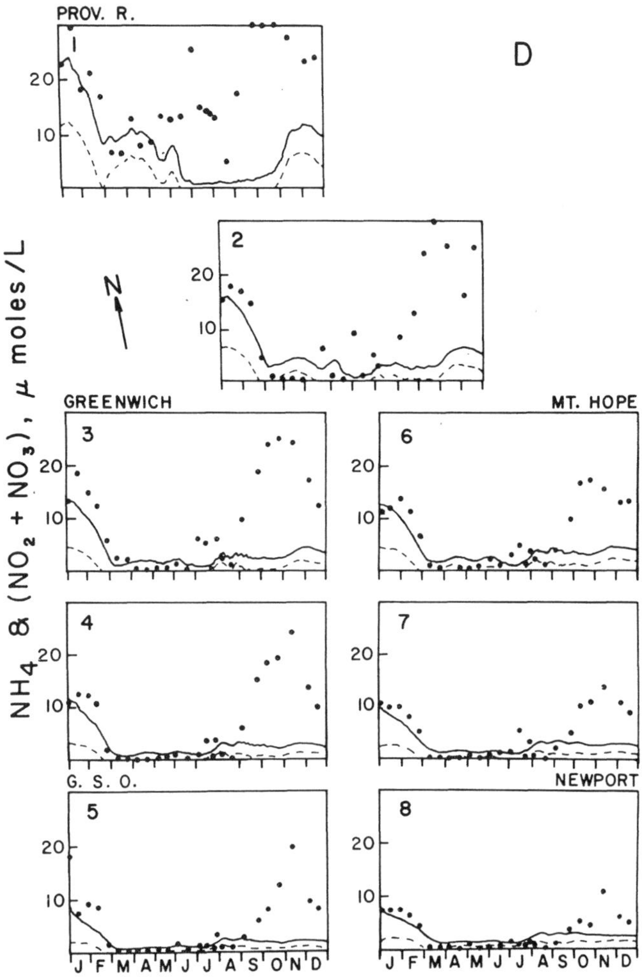

Fig. 40D (legend see page 136)

appropriate with observations of the midwinter minimum levels in Block Island Sound (Jeffries and Johnson, 1973), a particulate food subsidy of 0.1 mgC/l was made available to the copepods, and the dilution loss of adults by tidal flushing at the mouth of the bay was eliminated. Both of these adjustments are believed to be readily defendable—particulate carbon loads presumably utilizable to some degree by copepods far exceed this minimum (Oviatt and Nixon, 1975)—and, further, were only necessary during a short part of the simulation. With the onset of the spring phytoplankton bloom, zooplankton readily met their respiration and flushing losses, and the population responded appropriately. Figure 41 compares the standard run with the additions (dashed) to a run with no detritus or flushing changes, demonstrating how the maintenance of a low winter zooplankton population may be enhanced, thus permitting more rapid response to the

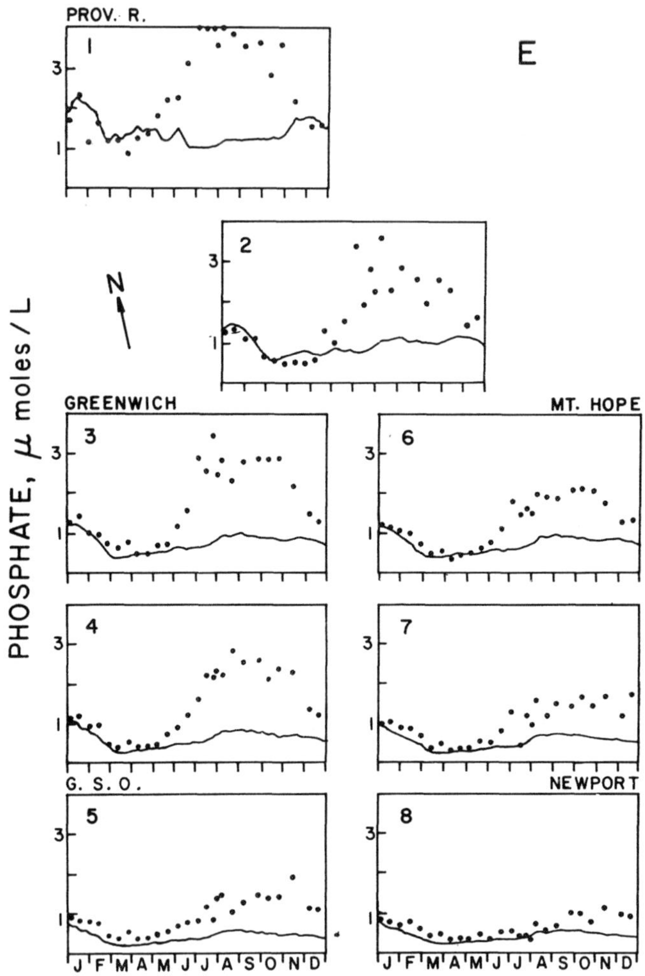

Fig. 40 E (legend see page 136)

phytoplankton bloom. The first generation of zooplankton in the spring appear as a small peak in April moving into the adult compartment in May, timing which approximately agrees with the observed peak, although nowhere near enough biomass is developed in the model. The summer period, as with the phytoplankton, is marked by oscillations suggesting relatively fast-turnover phytoplankton–herbivore cycles. Zooplankton biomass declines from late August on, though not as markedly as the observations, resulting in a consistent overestimate throughout the fall, in contrast to the general underestimate during the earlier seasons.

The seasonal and spatial distribution of silica predicted by the model agrees quite satisfactorily with the observed data (Fig. 40C). The decline commensurate with the winter bloom, the rapid recovery early in the summer, the gradual winter loss, as well as the spatial gradients around the bay are generally consistent. The

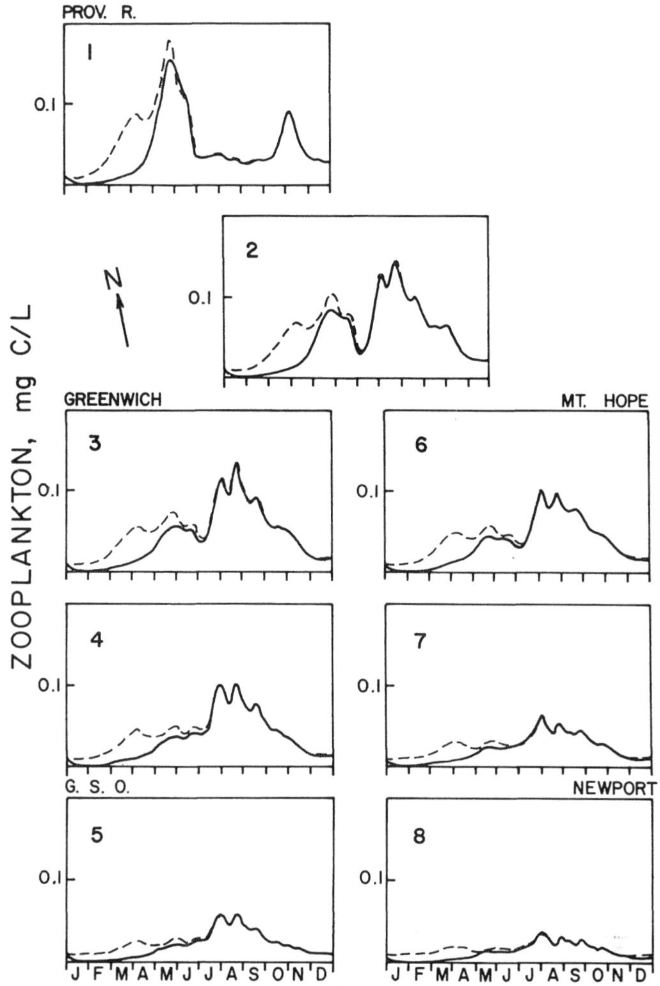

Fig. 41. Total zooplankton biomass resulting from removal of two features of the standard run designed to maintain low, stable zooplankton levels throughout the winter. When total biomass fell below 0.01 mg C/l in the lower bay, the standard run *(dashed)* routinely included a particulate carbon food subsidy and minimized flushing losses

sharp dip in silica during August throughout the bay is not portrayed in the simulation. This feature was apparently related to a pronounced chlorophyll peak occurring at the same time in the upper bay in August 1972, however it is not entirely explicable in this way, since the bloom was not baywide, and in most areas the silica minimum was not associated with any apparent flowering (Nixon et al., in preparation). In any case, the absence of this in the model simulation is to be expected, and the overall pattern is encouraging. It should be noted that this agreement was found to depend on a seasonal variation in the C:Si ratio of the phytoplankton community. In the model, the ratio is formulated to follow tem-

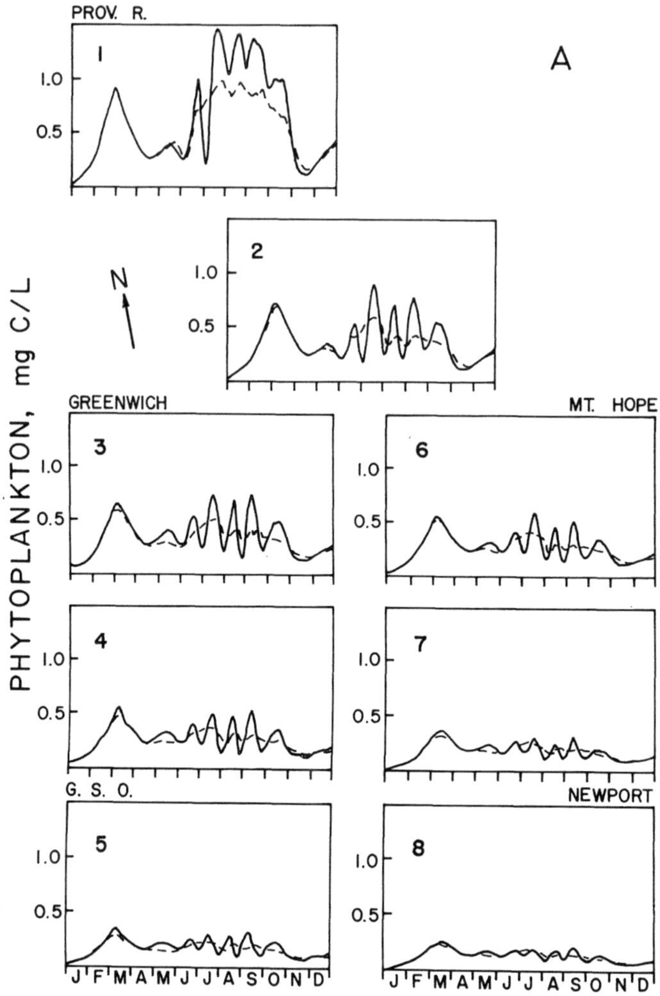

Fig. 42 A and B. Simulated phytoplankton (A) and zooplankton (B) resulting solely from a change in the algal nitrogen ratio from C:N = 7.0 in the standard run *(dashed line)* to C:N = 10.0 by atoms *(solid line)*

perature inversely, with the maximum silica requirement at the coldest temperature and vice versa. Despite the arbitrary nature of this scheme, it is a reasonable assumption in agreement with an observed temperature effect on diatom silica content (E. Durbin, personal communication). It also reflects the usual dominance of flagellates over diatoms during the warmer months.

Nitrogen and phosphorus patterns are poorly represented by the model in this standard run (Fig. 40D, E). A number of factors are believed to contribute to this disparity with some interesting implications for future research. The initial behavior of the model is satisfactory. Not surprisingly, the nutrients decline rapidly during the winter bloom and maintain appropriate low concentrations during the spring. In June, however, phosphorus begins an increase which is not explained by the

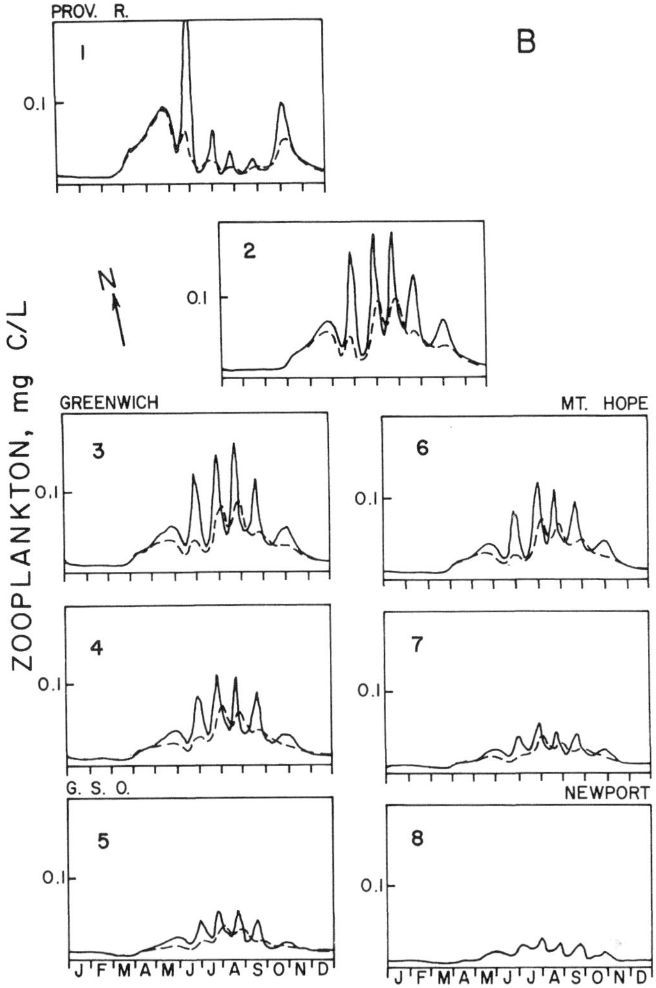

Fig. 42B

model. Nitrogen makes a temporary rise in July, attributable primarily to ammonia (Fig. 4A, B). And in September, nitrogen increases dramatically to its yearly maximum. The model increases also in July and August, but very slightly, and it maintains an unreasonably low level throughout the remainder of the year.

All the observed nutrients demonstrated a winter decline prior to the apparent phytoplankton bloom. Low temperature substantially reduces the benthic inputs, and simulations with the mixing model TIDE have suggested that the November decline may be due to flushing losses (Nixon et al., 1976).

A number of factors are believed to contribute to the inadequacy of the nitrogen and phosphorus formulations, as well as to the apparently low algal stocks, and additional simulations reveal some interesting considerations. Flagellate-dominated communities, which generally characterize the summer phytoplankton,

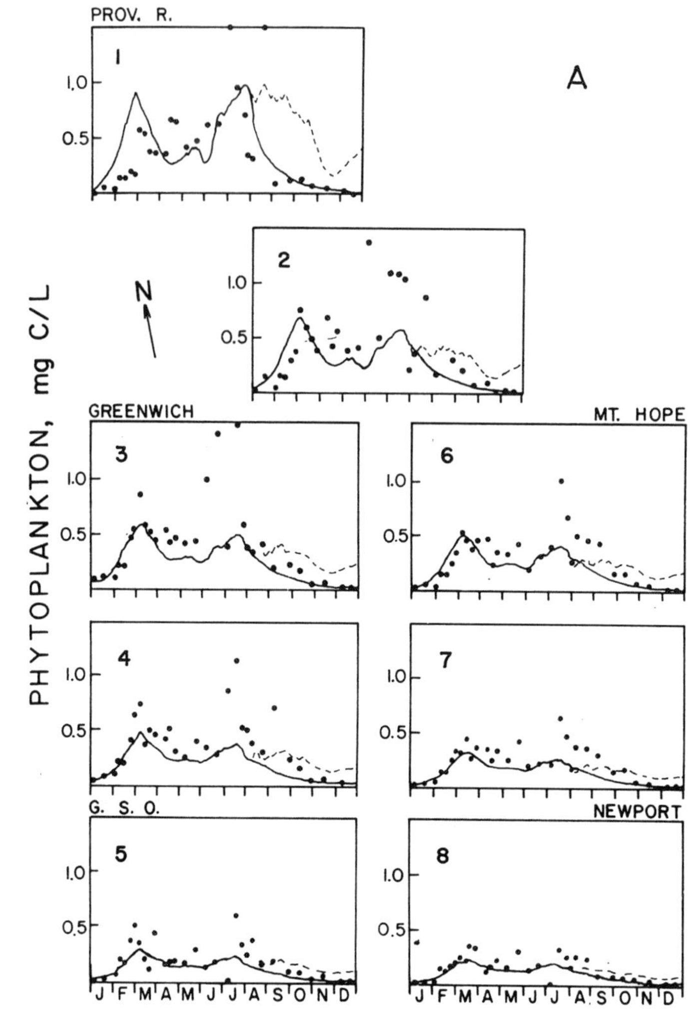

Fig. 43 A and B. Result of forcing an arbitrary 0.25% daily decline in the phytoplankton. Both phytoplankton (A) and total inorganic nitrogen (B) show improved agreement with observed data over the standard run *(dashed line)*

may require less nitrogen than the winter diatom species (Parsons et al., 1961; Strickland, 1966). When the constant C:N ratio for the algae is raised to 10.0 (by atoms) from the standard 7.0, the simulated phytoplankton differs markedly in the summer (Fig. 42A). The winter bloom and decline remain unaffected, since the onset of silica limitation prohibits further production. As the requirement for silica is lowered, the higher C:N results (1) in more-pronounced oscillations, and (2) in periods of increased standing stock. The fluctuations are related to herbivore grazing, as demonstrated by similar cycles in the zooplankton compartment (Fig. 42B). Copepod excretion is making recycled nitrogen available to the phytoplankton, and cannibalism during algal minima is reducing grazing pressure resulting in

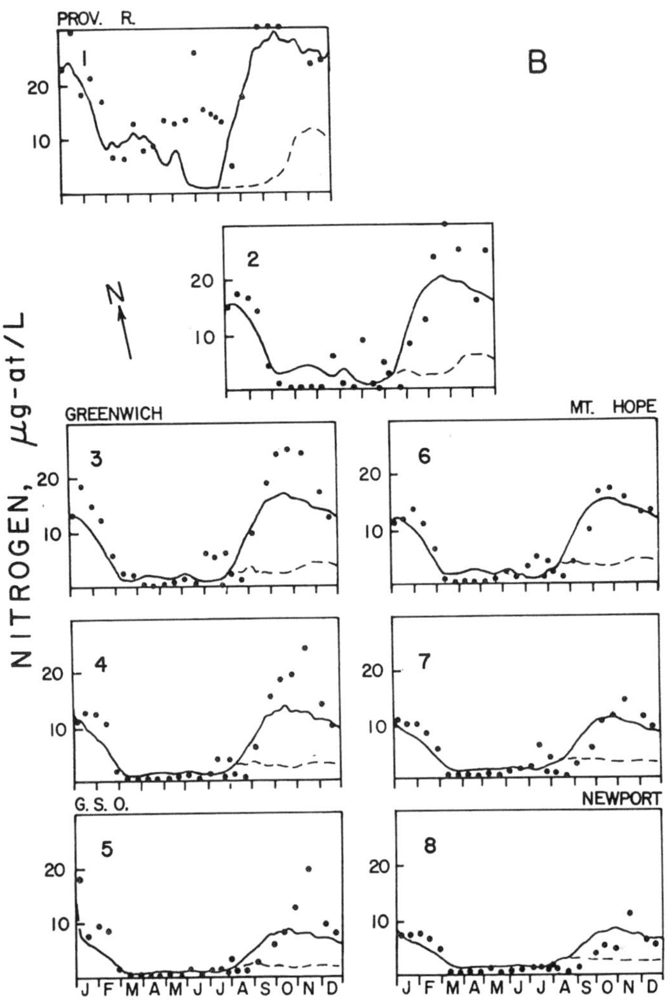

Fig. 43B

short bursts of relatively high-phytoplankton biomass. While such fast turnover dynamics probably approach the limit of reliability of the one-day integration step, some conclusions seem justified. First, the assumption of higher C:N requirements for the summer phytoplankton community may contribute to increased fluctuations in the nitrogen-poor environment as well as raising the peak biomasses observed during the oscillations. Second, in warm-water, fast-turnover situations, algae–herbivore cycles may intermittently move in and out of phase, increasing the likelihood of brief bursts of phytoplankton or zooplankton. In some model runs with different conditions this phenomenon was especially apparent, with single spikes of unusually high biomass resulting from a chance beating of the time-lagged cycles. While the model usually produces repeated, regular cycles due to its inherently tight coupling, the implications for natural systems are more general, and such processes may operate on much finer scales of patchiness, contributing to diverse heterogeneity in time and space.

Returning to the question of the summer behavior of the model, the previous paragraph presents possible explanations for the inability to develop appropriate summer biomasses with the available nitrogen in the system. The sensitivity to changes in the algal nitrogen composition implicates the total nitrogen formulation of the model. To evaluate the extent to which the pronounced observed phytoplankton decline might account for the late-summer nutrient rise, an arbitrary depression on growth was applied. In the increased C:N run, notice that most of the added primary production resulting from the change in the C:N ratio was converted to higher and more variable zooplankton biomasses due to enhanced animal metabolism at the warmer temperatures (Fig. 42). Benthic fluxes at this time are substantial, and nutrients are moving rapidly through the system even though they are not apparent in the concentrations. Thus, it may be anticipated that even small reductions in the standing crops will allow significant levels of nutrients to remain in the water. When the phytoplankton of the model were forced to decline a slight 0.25% daily from August 1, including all grazing demands, the depression in the algal biomass was small but significant when compared to the standard run (Fig. 43A). The consequences for the nitrogen concentration, however, were striking (Fig. 43B). The late summer and winter pattern was significantly improved, and both phosphorus and zooplankton agreements were also better.

An alternative scenario for the critical fall period is that the low biomasses are masking a highly active community of microplankton species. This is possible since microflagellates were noted as dominating the algal community (Nixon et al., in preparation), and the small zooplankton would not be adequately represented in the net samples. For this possibility, the nitrogen rise is less satisfactorily explained, since the microzooplankters would also tie up nutrients in biomass. Their high metabolic activity and short life span would perhaps result in such rapid regeneration that a buildup of inorganic nitrogen in the water might occur. And the model suggests that the potential for rapid turnover is certainly great. Productivity measurements for the fall of 1972 are quite low on a volume basis (Nixon et al., in preparation), and uncertainty in the carbon ratio for the measured chlorophyll makes a definitive conclusion tenuous on the possible turnover rate. Both scenarios remain plausible, and perhaps additional research will contribute to a better understanding of the critical late summer–fall period.

Regardless of the relative merits of the two explanations, the model run with the arbitrary phytoplankton decline suggests that the nitrogen and phosphorus entering the model system is insufficient to account for the total increase in inorganic concentrations. Urea-nitrogen is usually low (Vargo, 1976), but high concentrations of dissolved organic nitrogen are known to be regenerated by the benthos (Nixon et al., 1976) and have been measured in the water. Thus, a major component of the nitrogen budget may have been omitted from the model. Amino acids, or other dissolved forms directly or indirectly utilizable by the phytoplankton, may be playing a significant role in bolstering algal production during the summer–fall period, thus explaining the inability of the model to account for sufficient biomass. In the fall, as algal growth diminishes, the organic forms may be oxidized, releasing inorganic nitrogen into the water, in combination with continued fluxes from the sediments, accounting for the prominent yearly nutrient maximum. The absence in the model of a major feature of the annual nitrogen cycle for the bay is at the same time a disappointing shortcoming and a very useful result. It should be emphasized that data on the role of organic nutrients are presently insufficient to be evaluated in the model, and it is precisely such indications for continued research that the model was designed to provide.

9.5 Rates and Mechanisms

One of the most desirable features of a detailed mechanistic ecological model is that it enables detailed analysis not only of the emergent biomasses and concentrations of the system, but also of the many internal aspects of the dynamic processes. For any specified spatial regions, the bay model routinely prints out a set of values on the first of every month and the final day of the run. In addition, if more detailed resolution is required, weekly output may be specified for any period. An example of this output for three times during the standard run will demonstrate its usefulness (Fig. 44).

Characteristics of the physical forces in the bay are given in the first line. On 1 February, day 32, the temperature was approaching its minimum (TEMP = 2.8° C) in spatial element 4. The photoperiod was 9.7 h (DAYLT = 0.4 day), the average cloudiness for the past week (CVRAV) was seven-tenths, resulting in average insolation (RADIAV) of 131 ly/day. And the combined river flow entering the bay was 7.7 million cubic meters per day (QM3/D). For the other regions of the bay, only the temperature would be different.

The next set of variables pertains to the PHYTOPLANKTON compartment. The extinction coefficient (K) has increased due to the standing stock from the nonchlorophyll base of 0.36 (Table 7) to 0.53 m^{-1}. The optimum insolation for photosynthesis (IOPT) was 93 ly/day, but the asterisk reveals that the ambient insolation has been so low with respect to the required average for the water column (see REQIAV, LOIOPT, Chap. 4) that acclimation has been restricted. On a 24 h basis, the algal community was only able to grow at 16% of its maximum rate due to the integrated effects of light limitation (LTLIM). Ambient nutrients were still sufficient to meet the total demand of the algae, with the uptake: supply ratios well above unity (DLQN = 42, DLQP = 51, DLQSI = 48). The sinking rate was zero, so no loss to the benthos occurred (SINK = 0 mgC/l). The phytoplankton biomass (P1MGC = 0.13 mgC/l) was predicted by the temperature relationship [Eq. (7)] to have a potential maximum instantaneous 24 h growth rate of P1GMAX = 0.7 day^{-1}, or one doubling daily. The realized growth, however, was reduced by light and nutrient limitations to 13% daily net primary production throughout the water column (GP1 = 0.128 day^{-1}). Of the three inorganic nutrients, nitrogen was the most limiting, though it was sufficient to allow 84% of the light-limited growth (NLIM1 = 0.84, PLIM1 = 0.94, SILM1 = 0.96). The actual increase in carbon by the phytoplankton, the net community primary production, was DELP1 = 0.0161 mgC/l. MXLM1 repeats the most-limiting nutrient value, in this case MXLM1 = NLIM1 = 0.84. And, finally, the nutrient ratios for the phytoplankton community were C:N = 7, C:P = 85, and C:SI = 5.9, by atoms. In this run, nitrogen

A

```
DAY  32          MO    2  ELT  4                  TEMP =     2.7804  DAYLT=     0.4062  CVRAV=      7.0439  RADAV=  130.7320  QM3/D=0.7733E 07
PPL  K      =    0.5308   IOPT =      93.2439*  LTLIM=     0.1639  DLQN  =     41.59  DLQP  =      50.69  DLQSI=     48.06  SINK  =  0.0
     P1MGC=      0.1323   P1GMX=       0.7035   GP1   =    0.1282  NLIM1=     0.8398  PLIM1=      0.9407  SILM1=     0.9589  DELP1=  0.016133
     MXLM1=      0.8398                         C:N= 7.0                      C:P= 85.0                 C:SI  = 5.9
ZOO  ZTOT =0.9997E-02     ZAMGC=0.8226E-02      RTNMX=     0.2944  XRTN  =     0.6128  RLF   =     0.0112  XRESP=     0.1213  HATCH=    8
     +ADLT=0.0            ZJTOT=0.1771E-02      GROMX=     0.0396  XRTNJ=     0.6039  RLFJ  =     0.0023  XRSPJ=     0.1746  DEVEL=   76
     URA = 0.1536E-03     EGGS =0.1338E-02      GRO   =    0.9509  GTERM=     0.9403  ZJTMP=0.1098E-03  =>CAR=0.0
     EXC N=0.1563E-01     EXC P=0.1301E-02      C:N   =    7.0712  C:P   =    85.0027  N/P   =    12.0209  XN EX=     0.0938  XP EX=    0.1327
BEN  RTNP1=0.1717E-02     RTNP2=0.0             VFILT      1.5644  CGAIN=    12.1893  CREQ  =    24.4103  RLF   =    0.0136  SINKN=    0.0
     SINKP=     0.0       SNKSI=      0.0       SEDN =-.6472E 04   SEDP =-.1161E 04    SEDSI=-.3937E 05  FXNH4=0.4104E-01  FXPO4=0.5903E-02
     FLXSI=0.1941E 00
NUT  NH4  =      2.5774   NO2/3=      5.2882    PO4        0.7938  SI    =    11.6742
NET  %:  P1=    10.63 P2=    0.0  P =   10.63   AN=  -5.01 ON=   0.27 TN=  -1.52 PH=  -0.98 SI=  -0.42 ZJ=  -0.14 ZA=   0.0  TZ=  -0.03
```

B

```
DAY  213         MO    8  ELT  4                  TEMP =    20.3845  DAYLT=     0.5960  CVRAV=      6.0899  RADAV=  389.3447  QM3/D=0.1700E 07
PPL  K      =    0.6379   IOPT =     141.9886   LTLIM=     0.2381  DLQN  =      2.52  DLQP  =       8.40  DLQSI=     47.95  SINK  =  0.0
     P1MGC=      0.2275   P1GMX=       2.1441   GP1   =    0.3695  NLIM1=     0.6457  PLIM1=      0.9356  SILM1=     0.9783  DELP1=  0.084616
     MXLM1=      0.6457                         C:N= 7.0                      C:P= 85.0                 C:SI  =14.7
ZOO  ZTOT =0.1021E 00     ZAMGC=0.2308E-01      RTNMX=     0.8285  XRTN  =     0.8209  RLF   =     0.0650  XRESP=     0.4108  HATCH=    1
     +ADLT=0.0            ZJTOT=0.7898E-01      GROMX=     0.2304  XRTNJ=     0.7984  RLFJ  =     0.2743  XRSPJ=     0.5916  DEVEL=   20
     URA = 0.3083E-02     EGGS =0.5833E-02      GRO   =    1.0893  GTERM=     1.0101  ZJTMP=0.2423E-02  =>CAR=0.1145E-02
     EXC N=0.6478E 00     EXC P=0.5552E-01      C:N   =    7.2861  C:P   =    85.0180  N/P   =    11.6685  XN EX=     0.3808  XP EX=    0.5549
CARN CTENS=      0.0060   CTFLT=       0.0006   FISH  =    6.0000  FPREF=0.1102E-02  RTN   =C.1145E-02  FCARN=     0.0106  CTEXN=    0.0001
     MENH =0.1000E-04     M RTN=0.4815E-02      FMENH=0.3707E-01   IN ELEMENT 2
BEN  RTNP1=0.9987E-02     RTNP2=0.0             VFILT      5.0000  CGAIN=    70.9065  CREQ  =    78.0191  RLF   =    0.0436  SINKN=    0.0
     SINKP=     0.0       SNKSI=      0.0       SEDN =-.6893E 05   SEDP =-.1408E 05    SEDSI=-.4635E 06  FXNH4=0.5754E 00  FXPO4=0.5821E-01
     FLXSI=0.1346E 01
NUT  NH4  =      1.5188   NO2/3=      1.2503    PO4        0.7283  SI    =    22.5669
NET  %:  P1=    -1.34 P2=    0.0  P =   -1.34   AN=   7.81 ON=  10.60 TN=   9.05 PH=   4.52 SI=   4.13 ZJ=  10.26 ZA=  -2.12 TZ=   7.27
```

C

```
DAY  365         MO   12  ELT  4                  TEMP =     4.9520  DAYLT=     0.3768  CVRAV=      4.6436  RADAV=  144.8774  QM3/D=0.6191E 07
PPL  K      =    0.5912   IOPT =      93.8323*  LTLIM=     0.1483  DLQN  =     14.08  DLQP  =      35.03  DLQSI=     50.25  SINK  =  0.0
     P1MGC=      0.1893   P1GMX=       0.8072   GP1   =    0.0886  NLIM1=     0.6382  PLIM1=      0.9176  SILM1=     0.9582  DELP1=  0.016364
     MXLM1=      0.6382                         C:N= 7.0                      C:P= 85.0                 C:SI  = 6 4
ZOO  ZTOT =0.1505E-01     ZAMGC=0.9205E-02      RTNMX=     0.3345  XRTN  =     0.7508  RLF   =     0.0118  XRESP=     0.1410  HATCH=    6
     +ADLT=0.1054E-04     ZJTOT=0.5844E-02      GROMX=     0.0492  XRTNJ=     0.7344  RLFJ  =     0.0076  XRSPJ=     0.2030  DEVEL=   86
     URA = 0.5266E-03     EGGS =0.3063E-02      GRO   =    0.9822  GTERM=     0.9707  ZJTMP=0.3254E-03  =>CAR=0.0
     EXC N=0.2889E-01     EXC P=0.2499E-02      C:N   =    7.3576  C:P   =    85.0316  N/P   =    11.5570  XN EX=     0.1152  XP EX=    0.1694
BEN  RTNP1=0.3574E-02     RTNP2=0.0             VFILT      2.2189  CGAIN=    25.3731  CREQ  =    34.6228  RLF   =    0.0194  SINKN=    0.0
     SINKP=     0.0       SNKSI=      0.0       SEDN =-.2049E 06   SEDP =-.3349E 05    SEDSI=-.1053E 07  FXNH4=0.5684E-01  FXPO4=0.7829E-02
     FLXSI=0.2465E 00
NUT  NH4  =      0.8515   NO2/3=      1.7969    PO4        0.5569  SI    =    11.4620
NET  %:  P1=     5.12 P2=    0.0  P =    5.12   AN= -10.98 ON=   0.55 TN=  -3.47 PH=  -0.89 SI=   0.17 ZJ=   2.32 ZA=   0.11 TZ=   0.99
```

Fig. 44A–C. Sample output from the ecological model for three days during the standard run. See text for explanation and discussion

and phosphorus ratios were constant, but the changing silica ratio depicts the requirements of a cold-water, diatom-dominated population.

In the ZOOPLANKTON subroutine, the total biomass was about 0.01 mg C/l (ZTOT), of which most was reproductively mature adults (ZAMGC=0.008 mg C/l). The temperature-dependent maximum adult ration was 29.4% of the body carbon (RTNMX), but the available phytoplankton, eggs, juveniles and particulate food only permitted approximately 61% (XRTN) of this ration [Eq. (24)]. The combined adult biomass in 1 liter would have achieved this fraction of their maximum ration by filtering only 11.2 ml (RLF=.0112 $1\ l^{-1}\ day^{-1}$). This apparently low rate is largely due to the small biomass, for it represents almost 500 ml mg $dw^{-1}\ day^{-1}$. The weight-specific respiratory rate for adults was XRESP=0.12 day^{-1}. Adults were reproducing, and eggs produced today were to hatch 8 days later (HATCH=8 days). No new adults reached sexual maturity (+ADLT), probably because of the long development time at these temperatures (DEVEL=76 days). The juvenile biomass (ZJTOT=0.0018 mg C/l), was low, approximately 20% of the adults. The juvenile potential ration (see Chap. 5) is based on the temperature-dependent maximum growth (GROMAX=0.04 day^{-1}) and the weight-specific respiration (XRSPJ= 0.175 day^{-1}). The food-density limitation term (XRTNJ=0.604) reduces this projected ingestion resulting in the estimated filtration rate (RLFJ=0.0023 day^{-1}). The combined effects of cannibalism, low temperature and food limitation at this time kept the juvenile population from increasing. Although adults were able to meet their metabolic demands, with a positive unrespired assimilation going into egg production (URA=0.15×10^{-3} mg C/l), and had apparently been doing so for some time—the total eggs were EGGS=0.0013 mg C/l—the juveniles were not able to grow. Their ingestion provided for only 95% of the respiratory plus growth demands (GRO=0.95). And combined with predation from adult cannibalism, the net change in the juvenile compartments excluding tidal circulation effects was a 6% decline (GTERM=0.94). ZJTMP=0.11×10^{-3} mg C/l indicates the portion of the egg compartment that hatched into juveniles during this day, and none of the zooplankton biomass was lost to carnivorous grazing (=>CAR=0). The final values in this section concern nutrient regeneration by the total zooplankton compartment. The population was excreting 0.016 μg-at NH_4-N/l (EXC N) and 0.0013 μg-at PO_4-P/l (EXC P) daily into the water. The ratios of carbon respired to nutrients excreted were C:N=7.1 and C:P=85 by atoms, resulting in an excreted ratio of N:P=12. Finally, this excretion rate represented a daily loss of 9% and 13% of the body nitrogen and phosphorus respectively (XN EX=0.094 and XP EX=0.133).

In the BENTHOS compartment, the hard-clam population consumed 0.0017 mg C/l (RTNP1) of phytoplankton averaged throughout the water column. The temperature-dependent pumping rate was 1.56 l/h for each clam (VFILT). When this rate is prorated over 24 h for the local clam density (2.58 per m^2 in element 4, Table 10), the resultant filtering rate acting on the phytoplankton was RLF=0.0136 day^{-1}. The total consumption of the clams for the day was 12.2 mgC m^{-2} (CGAIN), in comparison to the calculated metabolic requirement of 24.4 mg C m^{-2} day^{-1} (CREQ) based on the pumping rate (Chap. 6). In this case the clams were able to meet only 50% of the estimated metabolic demand even assuming complete assimilation of the plankton in the filtered water. No nutrients were added to the sediments due to sinking (SINKN = SINKP = SINKSI = 0.0 μg-at/l), since for this run the algal sinking rate was assumed to be compensated by adequate vertical mixing. And the cumulative balance of nutrients for the benthic community during the first month of the simulation shows that more nutrients have been returned to the water than have reached the bottom by zooplankton fecal contributions and clam grazing (SEDN $= -6.5 \times 10^3$, SEDP $= -1.2 \times 10^3$, SEDSI $= -3.9 \times 10^4$ μg-at/m^2).

The second output list presented in Figure 44B depicts the model conditions on 1 August, day 213. The first feature of this series is the appearance of a new compartment. During the summer, the presence of carnivores in the model is indicated. This output is suppressed when no ctenophores, fish larvae, or menhaden are present in the bay. Since no menhaden are assumed to be feeding in element 4 (Chap. 6), a reminder of their role in element 2 of the upper bay is listed, including the biomass (MENH, pounds wet wt/l), preferred ration (MRTN, mg C/l), and filtering rate (FMENH, day^{-1}). For ctenophores, biomass (CTENS, mg dry wt/l), filtering rate (CTFLT, day^{-1}) and nitrogen excretion (CTEXN, μg-at N/l day^{-1}) are listed. And for larval fish, the values give the abundance (FISH, no./m^3) and their preferred daily ration (FPREF, mg C/l). The remaining two values combine the impact of all the carnivore components indicating the resultant total grazing pressure on the zooplankton population (FCARN $= 0.011$ day^{-1}) and the total net consumption of zooplankton by all carnivores (RTN, mg C/l). At this time, fish larvae were the dominant predators in the mid-West Passage, accounting for almost all of the 1.1% instantaneous rate of carnivorous grazing.

While detailed dissection of the dynamic picture portrayed by these output lists is cumbersome, it is these "snapshots" of the mechanisms underlying the simulated patterns that provide the feedback which is the most useful feature of the detailed ecosystem model. On the August day described above, a careful analysis reveals that the zooplankton community was actively feeding and more than meeting its metabolic demands for growth and reproduction, yet the combined predation of carnivorous zooplankton and adult cannibalism nearly exceeded the growth potential of the juveniles. Thus, food-limitation was relatively unimportant (XRTN and XRTNJ were about 80%), reproduction was occurring with a minimum generation time of 20 days, and juveniles were averaging a 9% daily realized growth (GRO = 1.089). Yet the combined predation reduced the final net juvenile change factor, GTERM, almost to 1.0, suggesting a significant degree of cannibalism.

The last output list (Fig. 44C) presents the state of the model on the last day of the simulation. The potential for phytoplankton growth is apparent, zooplankton were almost stable, with adults just exceeding respiratory demands (URA > 0), while juveniles were unable to do so (GRO = 0.98). The additional complement of newly hatched eggs (ZJTMP) balanced the metabolic death of the juveniles maintaining a slight net increase for the day. The sediment–nutrient terms (SEDN, etc.) indicate that a substantial imbalance is occurring in the model between the forced fluxes and the supply to the benthos. Over the long run, inputs of living or detrital material must provide sufficient additional nutrients to compensate for this discrepancy.

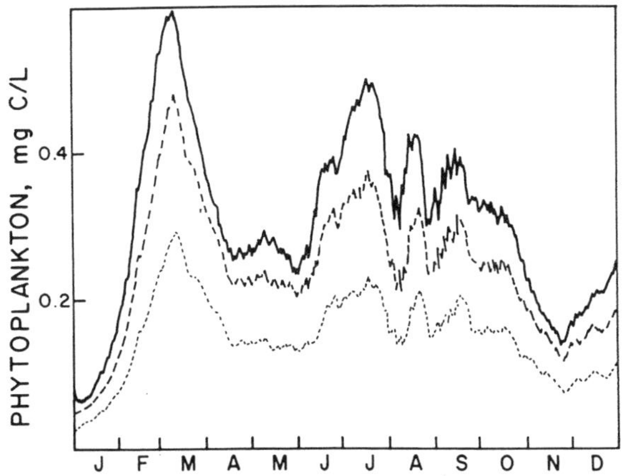

Fig. 45. Simulated gradients of phytoplankton down the West Passage, elements 3 *(top)*, 4 *(middle)*, and 5 *(bottom)*, for the standard run (see Fig. 1). The range of concentration is decreased and the inception of the winter bloom is slightly delayed downbay

A number of further observations on the general indications of the standard run are possible. The potential role of depth in influencing the early initiation of the winter phytoplankton bloom has been offered as a partial explanation for its distinctive timing (Pratt, 1959; Smayda, 1973a). Evidence supporting this hypothesis includes the observation that the bloom appears to begin earlier in the shallower upper reaches of the bay. Some support for this view is evident in the model results. Figure 45 is a comparison of the standard simulations in the three spatial elements of the West Passage. For the range of average depths in these elements (4.6, 7.1, and 9.9 m, respectively) the apparent delay of the bloom maximum is slight. On each day, however, the clear differences in magnitude effectively represent the bloom as occurring earlier in the upper bay. That is, any given biomass is reached in element 3 a week or two ahead of the lower regions. In the model this acceleration results from a combination of factors. The temperature is slightly higher in the upper bay in February and increasingly so in March (Table 1), so the potential maximum growth is greater. Nutrients show a downbay gradient, though generally high, and have some relative effect. And the depth and self-shading differences are mediated through the light limitation term. The interactions resulting in the instantaneous growth estimates may be summarized for February 1 in the three elements:

	Temp °C	GMAX	NLIM	LTLIM	GP1 day^{-1}
Element 3	2.78	0.7035	0.8674	0.2138	0.1568
Element 4	2.78	0.7035	0.8398	0.1639	0.1282
Element 5	2.58	0.6947	0.7611	0.1331	0.0992

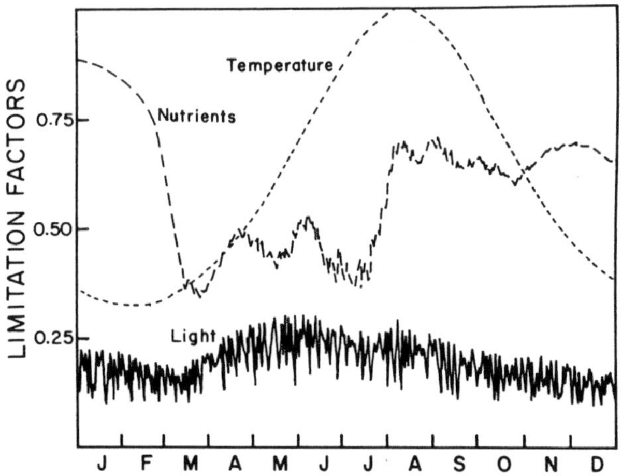

Fig. 46. The relative seasonal role of temperature, light, and nutrients in affecting phytoplankton growth. Temperature defines the potential rate, here plotted relative to the seasonal maximum, which is subsequently reduced by unitless limiting factors based on nutrients and light

The daily growth rate, GP1, is calculated in the model as the product of GMAX, NLIM and LTLIM, as has been discussed in Chapter 4. In this example, it should be noted that this product does not agree with the listed growth. This is due to the predictor–corrector integration scheme (see Chap. 7) which averages two estimates of GP1, while the printed values of the limitation terms represent only the corrector-step conditions.

The model can also provide a graphical synopsis of the relative role of the three limitation factors during an annual phytoplankton cycle (Fig. 46). While not used in this form in the model, the seasonal role of temperature in fixing the maximum growth rate can be expressed as a unitless ratio based on the midsummer yearly maximum. In this form, the range of the temperature effect is shown to be comparable to the regulation imposed by nutrient variations throughout the year. The NUTLIM curve depicts the onset of nutrient limitation associated with the winter–spring phytoplankton bloom, as well as the fluctuating low levels that follow. Light is consistently the most severe limitation, since the effect is integrated over 24 h and the entire water column. Yet, while light may consistently be the factor in shortest supply, the range of its seasonal regulatory role is less than that of either nutrients or temperature for the algal component of the system.

A detailed inspection of the relative importance of the three nutrients (Fig. 47) indicates that nitrogen remained the single, most-limiting nutrient throughout the simulation. Although phosphorus never approached limiting levels, silica depletion was comparable to that of nitrogen at the termination of the winter–spring bloom. In other simulations with different choices of half-saturation constants, silica replaced nitrogen as most limiting during this critical time. The relative effect of these two nutrients appears to be delicately balanced, and the contention by Pratt (1965) that silica plays an important role is supported by these results.

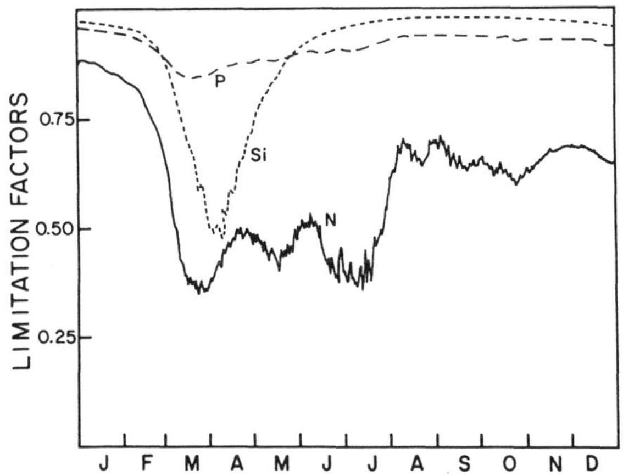

Fig. 47. Comparison of limitation of simulated phytoplankton growth due to nitrogen (N), phosphorus (P), and silica (Si). In this simulation, nitrogen was the most limiting nutrient throughout the year. In other runs, silica depletion rather than nitrogen appeared to terminate the winter-spring bloom

A prominent feature of the temporal patterns shown earlier in Figure 45 is the frequency of small-scale perturbations, especially during the summer. This "chatter" results principally from the stochastic daily variation of insolation. The effect is clearly demonstrated again by the baywide graphs of the instantaneous daily growth rates (Fig. 48). The complexity of factors interacting in the model is reflected in this critical rate of production. During the winter, insolation is low, and acclimation of the optimum light, I_{opt}, is inhibited by the arbitrary lower limit (Sec. 4.3). The daily variations of insolation have less effect under this essentially constant physiological response than in the summer when the extremes of absolute insolation are wider, in addition to a constantly changing I_{opt}. The widely fluctuating rate in the summer is especially interesting. With the random insolation regime of the model, the acclimating I_{opt} in some cases results in even less production than a constant average value. In nature, where cloudy or clear days may tend to occur together, a regular acclimation response would perhaps be a more satisfactory strategy. But the general indication of the model is that the tendency to acclimate to a basically unpredictable factor may result in a less-satisfactory productive capacity. Whether this theoretically derived conclusion has reasonable implications for the extension of controlled light regime acclimation studies to the natural in situ phytoplankton response is difficult to say from the model alone. But these results demonstrate that complex variability in systems may sometimes alter conclusions drawn from more simple experimental or conceptual premises.

Other factors besides the stochastic cloudiness contribute to the large variability in the daily growth rate. In times of nutrient depletion, as is the case in March (Fig. 46), the relative amplitude of the erratic changes is dramatically reduced (Fig. 48). Conversely, during the summer season of peak activity, oscillations in zooplankton biomass result in variable contributions of recycled nutrients, as well as periodic

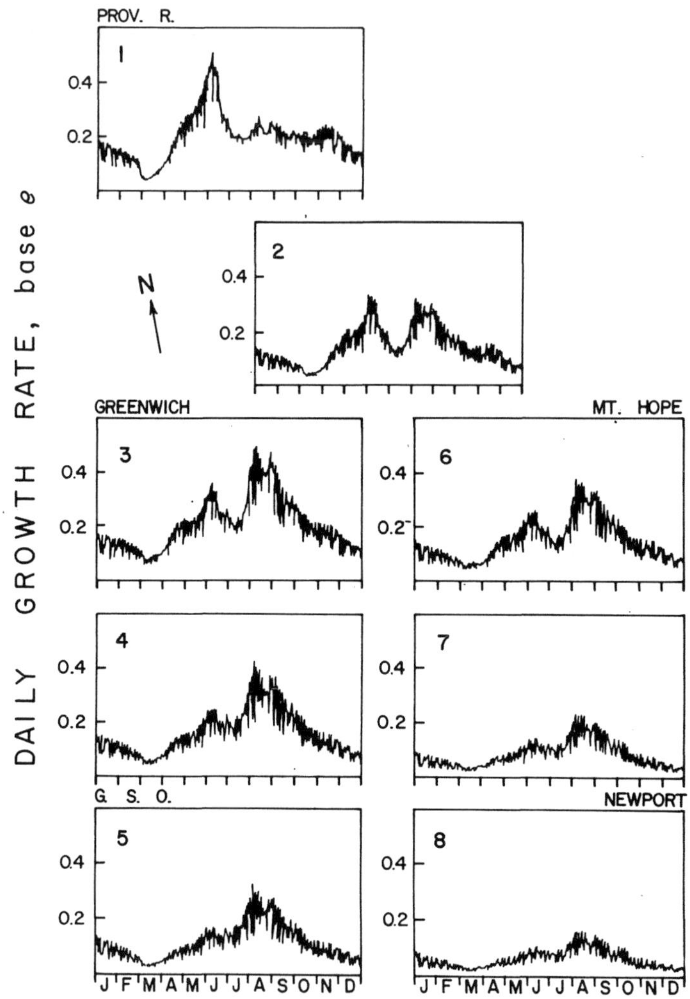

Fig. 48. Baywide patterns of the instantaneous daily growth rate (base e) of phytoplankton calculated in the standard run of the model

grazing intensity. Thus, the rapid algal biomass changes accompanying the erratic growth rates also reflect the impact of the herbivores.

9.6 Relative Role of Nutrients and Grazing in Phytoplankton Control

Because of the detailed analysis possible in mechanistic simulations, the model may be used to investigate specific hypotheses about the ecosystem. For example, a classical discussion concerns the relative importance of nutrient limitation and grazing in controlling phytoplankton abundance (see Cushing, 1964, 1968). In fact, Smayda (1973a) noted that this was still an open question concerning the termination of the winter–spring bloom in Narragansett Bay.

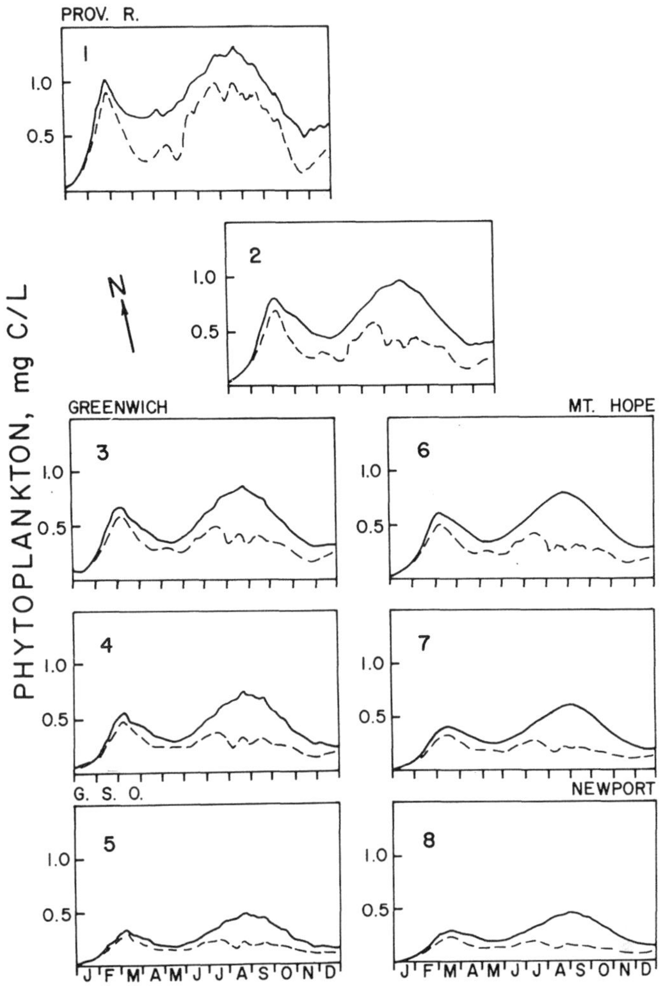

Fig. 49. Seasonal pattern of phytoplankton calculated in the model when zooplankton were removed from the system. The standard run is also given for comparison *(dashed line)*

When the bay model was run with zooplankton entirely removed from the system, the seasonal pattern of phytoplankton was remarkably similar to the standard run (Fig. 49). The major differences were a slight increase in the magnitude of the winter bloom, substantially higher summer biomass, and the lack of marked oscillations. The basic bimodal pattern represents a reasonable envelope within which most seasonal observations occur. Since all zooplankton grazing has been removed from the system, interactions related to nutrients, benthic grazing and flushing account for the gradual decline after the blooms. In the spring, benthic grazing, which has been formulated as a function of temperature, is gradually increasing, but accounts for a loss of from less than 1–4% daily (1.4–43.6 ml cleared per liter per day on February 1 and May 1, respectively, in element 4). Nutrients, of

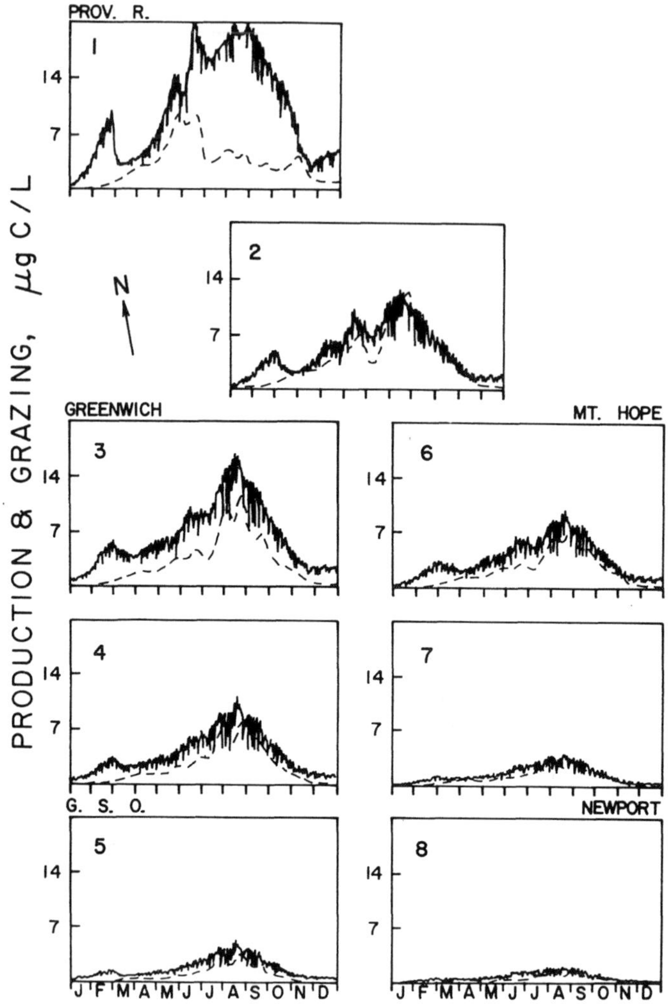

Fig. 50. Comparison of phytoplankton carbon production *(solid line)* and realized zooplankton ration *(dashed line)* for standard run

course, are at the annual minimum and this reduction in growth enables flushing losses to decrease the algal stock.

The above example is a very simplified case, but serves to demonstrate that appropriate phytoplankton declines in Narragansett Bay may potentially be explained without invoking grazing. A more complete picture results from a detailed look at the relative rates of phytoplankton production and herbivorous grazing for the standard run (Fig. 50). It is apparent that the daily production of carbon shows a depression due to nutrient limitation in early March, while copepod grazing is still at a relatively low level. In the summer, by contrast, zooplankton ingestion removes most of each day's primary production, severely limiting phytoplankton population increases. Thus, according to the model, nutrient limitation appears to be the prime factor in terminating the winter–spring bloom. Nutrients, however, can only reduce growth, and the actual decline may be due to flushing, plus some grazing by zooplankton and clams. The model does not include meroplankton, and as noted in the discussion of the observed data, a zooplankton peak in February 1973 corresponded approximately with a sharp phytoplankton decline. Therefore, the possibility remains that herbivorous grazing may in some instances play a role in the post-bloom decline, even though the copepod biomass may be small.

9.7 Zooplankton Excretion and Benthic Fluxes

Another question of interest is the importance of various nutrient sources in supplying phytoplankton assimilation. For the conditions of the standard run, algal uptake has been compared with two major nitrogen contributors in the bay model, benthic flux and copepod excretion (Fig. 51A and B). Throughout most of the simulation, both sources contribute about equally toward meeting the algal growth demands. During the coldest months, benthic fluxes remain slightly higher. This would be expected to be especially true during years of extremely low fall zooplankton abundance, such as 1972 (Fig. 6). At these times, however, carnivore excretion may become significant, at least in local regions of high density such as have been observed for ctenophores (P. Kremer, 1975a) and menhaden (Oviatt et al., 1972).

An additional point should be made. While the model attempted to represent the major features of the nutrient balance in the bay, its shortcomings in this regard have already been discussed. Thus, while it is believed that the relative nutrient fluxes presented here are reasonable, some additional factor is apparently missing. Additional research is essential, especially during the critical fall period, to lend further support or contradiction to these conclusions.

9.8 Metabolic Carbon Budgets for Zooplankton

While the routine output of the program provides monthly views of the conditions underlying the computed patterns, a more complete view is also valuable. For the standard run, the parameters critical to the zooplankton compartment were plotted daily to provide a synoptic view of the seasonal metabolic balance (Fig. 52). The weight-specific respiration rates are smooth functions of temperature [Eq. (28)], with the higher juvenile rate due to allometry [Eq. (30)]. The ingested ration is also primarily driven by temperature [Eqs. (23) and (34)], although substantial irregularities result due to changes in food availability and competition. For example, the late-summer oscillations are reflections of reduced phytoplankton stocks during the algal–herbivore cycles which characterize this period in the model. Assimilation is simply a constant fraction of ingestion, here 80%, but it is the relative magnitude of assimilation and respiration which are crucial to the status of the zooplankton compartment. For the adults, a positive net unrespired assimilation represents reproduction, and the

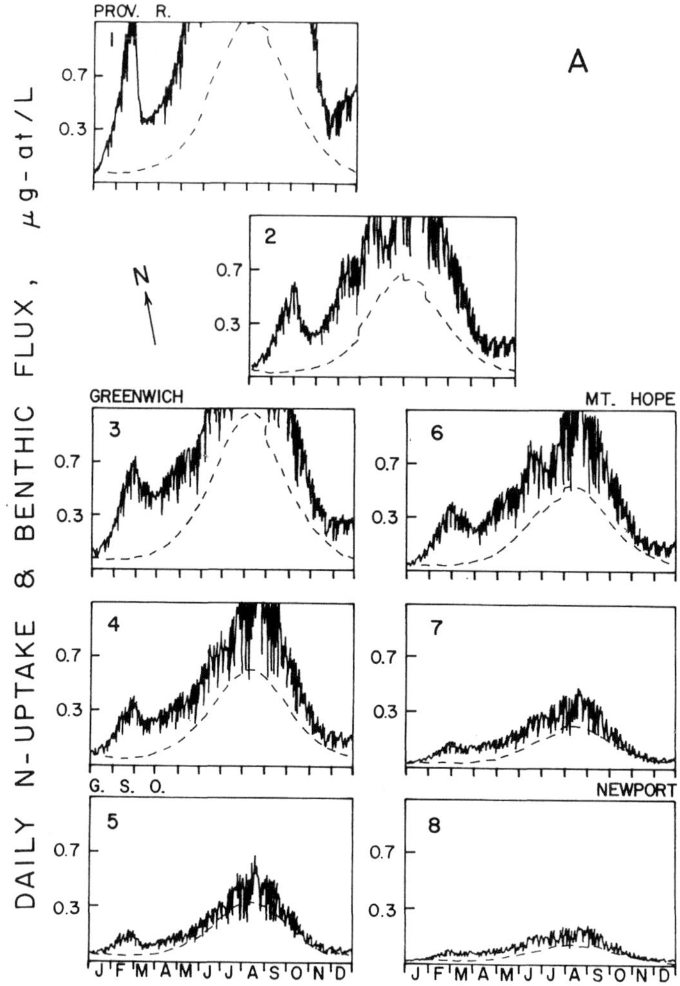

Fig. 51 A and B. Daily uptake of nitrogen by phytoplankton *(solid line)* compared with contributions of benthic fluxes (A) and zooplankton excretion (B)

model indicates that the population is producing eggs during most of the year. For the juveniles, however, the margin represents growth toward reproductive maturity, and their increased respiratory rate makes the balance marginal at many times during the annual cycle. Further, the depressed ingestion of the juveniles during the summer is more pronounced, since adults may supplement their ration by cannibalism, leaving the juveniles more susceptible to low phytoplankton levels.

Throughout the year, an integrated balance of the juvenile compartments' recent history is contained in the distribution of biomass in various stages of development. A detailed look at the simulated age structure of the juveniles (Fig. 53) reveals interesting features of the population dynamics. The initial reproductive cohort is distinctly obvious as it grows to maturity. As the season progresses,

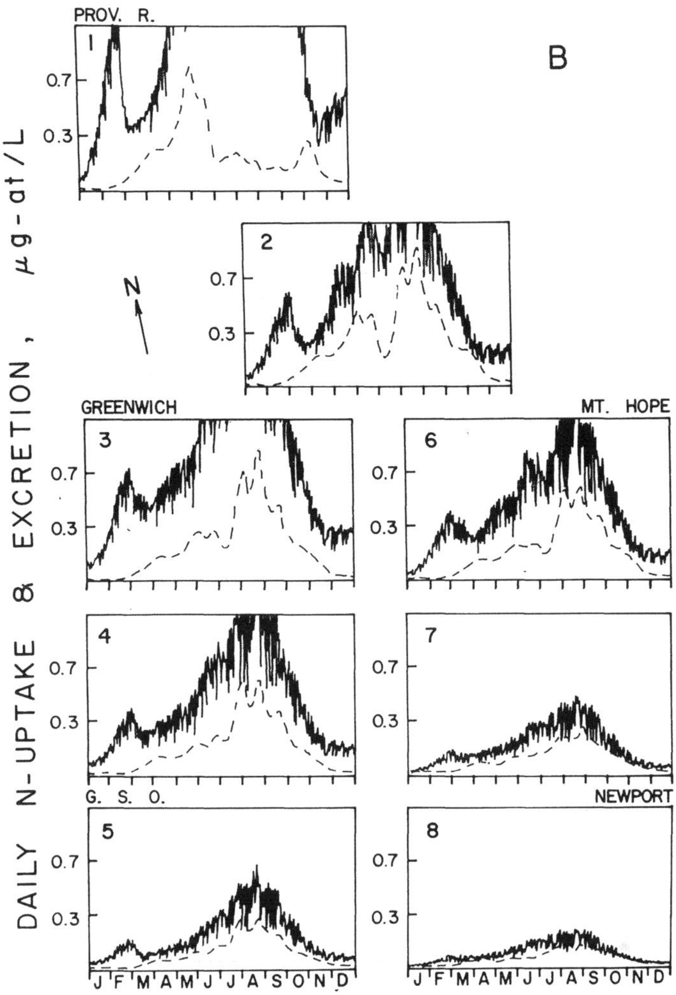

Fig. 51 B

however, the situation becomes confused, with shortened development times and overlapping generations obscuring any regular patterns. During the summer, the relative abundance of different developmental stages primarily reflects their unequal duration (p. 77), and any age-structure analysis would be inconclusive since the assumption of steady state is clearly unjustified.

9.9 Annual Integrals

Perhaps the best summary for the general results of the standard run is an annual integration of the carbon and nutrient fluxes calculated by the model. For

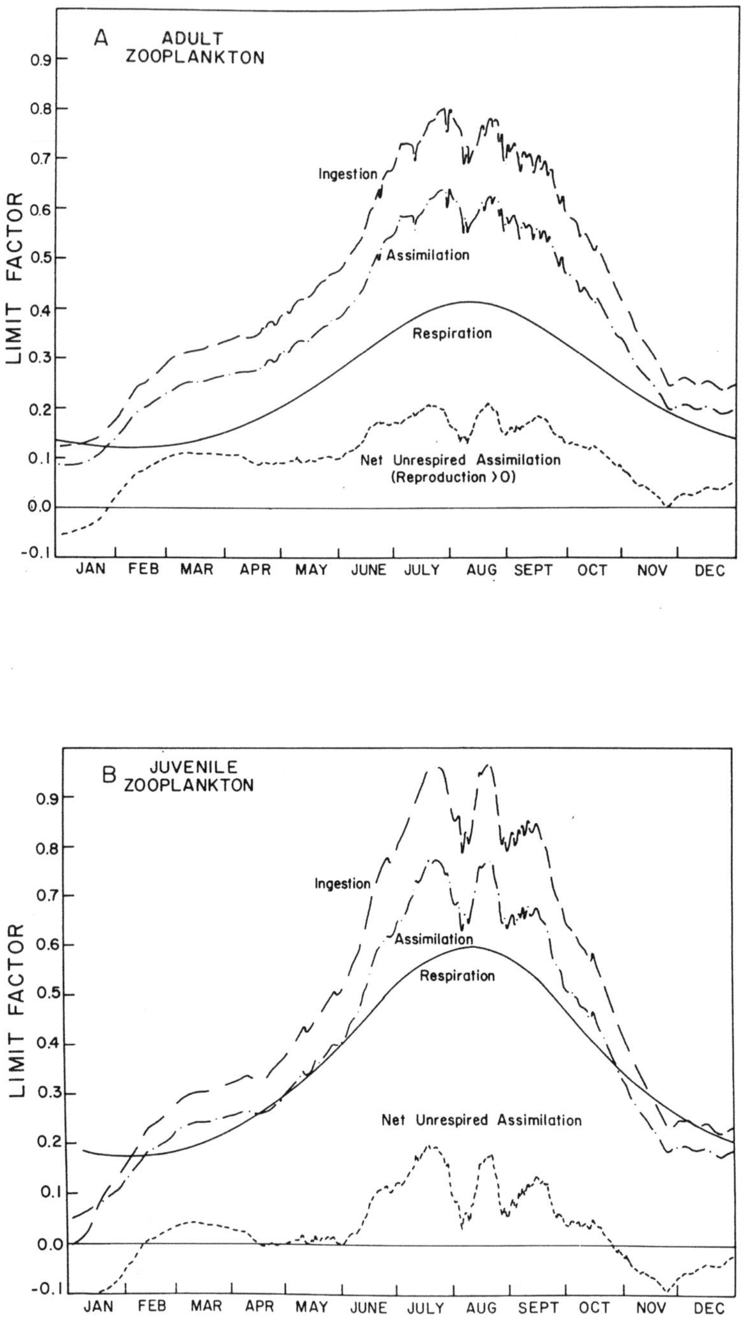

Fig. 52A and B. Simulated metabolic balance of zooplankton. Weight-specific daily rates of ingestion, assimilation, respiration and net unrespired assimilation are shown as limitation ratios vs. the temperature-dependent maximum ration for adults (A) and juveniles (B) for the conditions in element 4 of the standard run

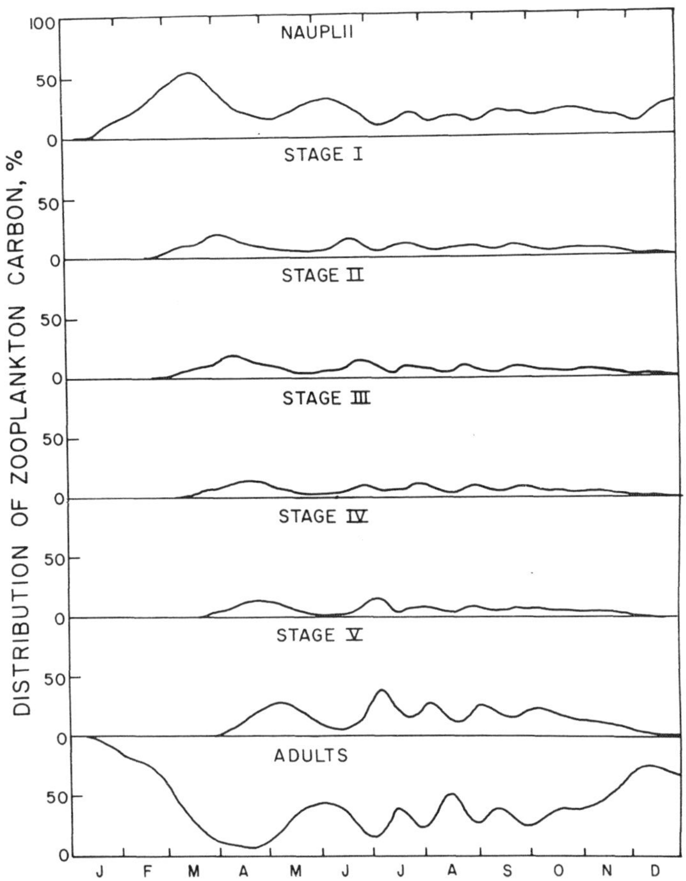

Fig. 53. Simulated age structure of juvenile zooplankton. The current maximum development time throughout the year was divided in proportion to the duration of stages for juvenile *Acartia clausii* according to the data of Petipa (1966; see p. 77). Graphs represent the total carbon biomass present in each of these divisions

each spatial element of the bay, the 365 daily changes were accumulated to provide a synoptic view of the relative budgets within the model (Table 13).

A number of these values deserve additional comment. The net annual primary production estimates (row 1) vary considerably around the bay, although the high Providence River value is probably extreme. The general range of 60–125 g C m^{-2} $year^{-1}$ seems low with respect to Smayda's estimate based on surface rates of 220 g C m^{-2} $year^{-1}$ (1957). However, our computed values take into account reduced production lower in the water column. A preliminary analysis of representative vertical productivity profiles for Narragansett Bay suggests the rates in the surface meter are often two to three times the average for the water column. In addition, there are a number of large transient blooms that occur during the summer that are not well simulated by the standard run.

In general, the bay seems to utilize the majority of both primary and secondary production, probably resulting in little export to Rhode Island Sound and nearby

Table 13. Summary of annual integration of carbon and nutrient fluxes for the standard run of bay model. Carbon estimates are g $C/m^2 \cdot$ year, and nutrients are mg-at/$m^2 \cdot$ year

			West Passage			East Passage		
Annual flux — Element:	Prov. R. 1	Upper Bay 2	Upper 3	Mid 4	Lower 5	Upper 6	Mid 7	Lower 8
1. Net primary production	154.8	127.3	105.3	97.1	59.2	102.3	77.3	61.1
2. Community primary production	45.3	−4.9	24.0	18.4	10.9	12.7	2.3	11.8
3. Net secondary production (copepods)	11.8	16.0	7.6	6.6	2.0	7.7	3.5	1.4
4. Community secondary production	−1.7	7.7	2.2	4.0	−1.8	2.9	−0.5	−5.8
5. Phytoplankton consumed by:								
a) Copepods	48.5	95.7	51.3	57.3	36.5	65.2	63.7	46.5
b) Benthos (clams)	61.0	36.5	30.0	21.5	11.9	24.4	11.4	2.7
6. Zooplankton consumed by carnivores	9.1	7.8	5.2	2.4	1.3	4.3	2.0	0.8
7. Benthic nutrient flux								
a) NH_4-N	633.6	646.5	622.3	601.8	530.4	584.8	544.4	463.1
b) PO_4-P	69.0	70.0	67.9	66.1	59.4	64.5	60.7	53.2
c) $Si(OH)_4$-Si	1732.6	1752.5	1709.9	1674.3	1534.1	1641.7	1561.8	1408.2
8. Sewage input								
a) NH_4-N	1095.5	50.6	25.2	20.7	0.0	85.4	0.0	549.6
b) NO_2 & NO_3-N	329.6	8.0	16.8	13.0	6.0	17.1	0.0	61.1
c) PO_4-P	146.0	8.0	6.7	2.6	0.0	11.4	0.0	52.3
d) $Si(OH)_4$-Si	23.4	1.3	0.2	0.5	0.0	2.6	0.0	7.0
9. Zooplankton excretion								
a) NH_4-N	313.1	676.4	366.5	451.1	360.9	506.5	579.9	523.4
b) PO_4-P	29.3	61.2	33.1	39.4	29.4	44.4	48.4	41.2
10. Ctenophore excretion NH_4-N	10.9	2.8	2.9	2.9	3.7	1.0	4.0	2.1
11. Uptake by phytoplankton								
a) Nitrogen	1843.0	1515.8	1253.5	1156.2	705.0	1217.3	920.8	727.4
b) Phosphorus	151.8	124.8	103.2	95.2	58.1	100.2	75.8	59.9
c) Silica	1229.1	1015.4	859.5	792.7	502.1	854.1	650.8	554.2
Depth (m)	4.3	7.3	4.6	7.1	9.9	7.8	15.2	23.9

coastal waters. Zooplankton and benthic grazing consume most of the production, with net community primary production averaging only 15% of the total annual production. The carnivores included in the model effectively consume most of the zooplankton, with only a small secondary production remaining after the one-year cycle. Interestingly, the regions of the bay which actually show a negative zooplankton balance are the Providence River, due primarily to the impact of menhaden, and the lower East and West Passages, due to flushing losses to Rhode Island Sound.

The nutrient budget reflects the relative importance of benthic fluxes and zooplankton excretion that were discussed earlier. An additional compartment which is poorly represented in the present model is the excretion of carnivores. Satisfactory data were only available for ctenophores, and their role is included for comparison. As better information becomes available on the other consumers, their addition to the model will improve the scope of the hypothetical budget presented here.

10. The Role of Biological Detail

One of the primary goals of this model was to include a high degree of detail in the phytoplankton and zooplankton compartments. It is therefore of particular interest to investigate the consequences of these additions in terms of possible implications for the natural ecosystem as well as for future modeling efforts.

10.1 Light Optima and Acclimation

The wide variations in the response of photosynthetic organisms to different or changing light intensities has been an area of intensive investigation for many years (see Sect. 4.3). For this reason, the model maintained wide flexibility in this formulation. Three options for inputing the daily insolation were provided, and the physiological response was treated in fairly sophisticated detail, including a scheme for light acclimation by the algal community.

For the biological conditions of the standard run—i.e., including the acclimation feature with a reasonable lower limit (Sect. 4.3.3)—the phytoplankton response to the options for solar input are compared in Figure 54. The solid line is

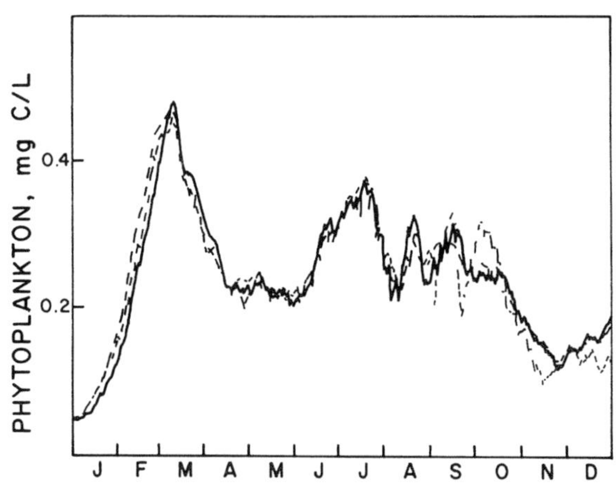

Fig. 54. Response of the simulated phytoplankton to three options for insolation: a stochastic cloudiness scheme *(solid line)*, a smoothed average cloudiness pattern *(longer dashes)* and actual observed data for Aug. 1972–Aug. 1973 *(finely dashed)*

the standard run which utilized the numerical stochastic cloudiness generator of the model. This scheme reproduces day-to-day variations in cloudiness with an appropriate seasonal distribution, and reduces the theoretical clear-sky maximum insolation accordingly [Eq. (5)]. The second line with the longer dashes used a smooth curve of average insolation with no daily irregularity, and the finely dashed line was run with actual data for the year August 1972–August 1973 (available through the Eppley Laboratory, Newport, RI). The differences in the three simulations are not striking. In fact, throughout most of the year the patterns are indistinguishable due to the stochastic variability. The model is reasonably stable with respect to these differences, probably due to the acclimation feature of the algae.

Two areas of difference are evident, however. First, the model's stochastic option predicted a slightly slower winter bloom than either of the others. The smoothed pattern was the fastest, as the acclimation scheme is most effective in optimizing growth potential during periods of regular insolation. The actual-data simulation was intermediate, perhaps reflecting a tendency for longer sequences of cloudy and light days than the stochastic scheme produced.

A second difference in the simulations is apparent during the fall. Hitchcock and Smayda (1977) have noted that the weekly averages of insolation during December 1972 were unusually low. The response of the model indicates this effect (finely dashed line). Both the general stochastic model and the smoothed pattern demonstrated a consistent slow increase from the middle of November, while the actual insolation resulted in an erratic but stable biomass in this period. Interestingly, during early October, the reverse was true. The actual data predicted a slight peak that was not apparent in the other two simulations. Thus, while the model substantiates the conclusion that low insolation may have played a role in limiting phytoplankton growth during the late fall, the earlier decline throughout the fall seems to have occurred despite more satisfactory insolation levels.

A second set of simulations further demonstrates the role of optimal light acclimation in the model. Figure 55 contrasts the standard run (REQIAV = 40 ly/day, solid line) with the cases of unrestricted (REQIAV = 0, dashed line) and severely restricted acclimation (REQIAV = 100 ly/day, finely dashed line). In the latter case, the optimal light for photosynthesis, I_{opt}, was inhibited from moving below the equivalent of an average insolation for the water column of 100 ly/day (see Sect. 9.3). For clarity, the stochastic variability has been removed from the two modified runs, and a slight difference is therefore to be expected (see Fig. 54). Even so, the effects of the altered physiological capacity were quite dramatic. As expected, the freely acclimating community was able to enter the bloom phase considerably faster than the standard run, since the low levels of insolation in January were below the reasonable acclimation threshold of 40 ly/day average insolation. Under extreme limitation, the bloom was delayed a full month, with a stable biomass predicted during January. Similarly, at the end of the year, severe light limitation reduced production sufficiently to suppress the November bloom inception shown in the other two runs.

While these model results simulate only the role of light inhibition of photosynthesis per se, in nature this physiological response may be affected by other factors. Thus, if the photosynthetic capacity is directly altered by nutrient

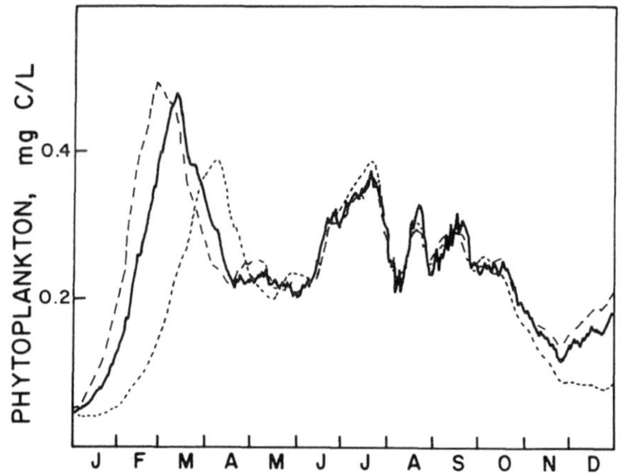

Fig. 55. Effect in the model of changes in the arbitrary lower limit to acclimation. Standard run (REQIAV=40, *solid line*) was compared with simulations of no lower limit (REQIAV=0, *dashed*) and an unusually high lower limit (REQIAV=100, *finely dashed*). Stochastic insolation removed from the two modified runs for graphical clarity

limitation, for example, dramatic changes in productivity similar to those discussed here might be mediated through the light response.

In the theoretical discussion of the photosynthesis–light response (Sect. 4.3), an argument was presented which concluded that the diurnal pattern of irradiance, combined with the vertical integration throughout the water column, might be expected to dampen the total daily response due to small changes in I_{opt}. A critical consideration was whether the photosynthetic optimum light occurs within the

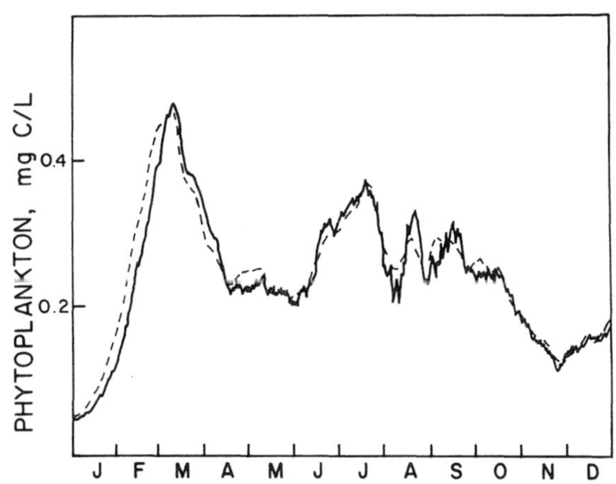

Fig. 56. Comparison of the standard run including light acclimation *(solid line)* with a run assuming a low, constant I_{opt}. Note the lack of significant differences during the bright summer months despite severe surface inhibition in the unacclimating case. Stochastic insolation removed from the modified run for graphical clarity accounts for most of the difference (see Fig. 54)

water column. That is, if the incident radiation at the surface is less than the optimum, the total water column is subject to limitation, and further decreases in light will reduce the production of the community. If the optimum occurs at some intermediate depth, however, changes in the light level only shift the site of optimum production, and in general the total production is less affected. In short, the effect of surface inhibition, demonstrable in the instantaneous rate response in constant light, is less critical in affecting the production of the community when effects of day-length and depth are considered. This conclusion is also demonstrated in Figure 56, which compares the standard run with dynamic acclimation to a simulation with constant low I_{opt}. Throughout the year, I_{opt} was defined by the lower limit to acclimation, REQIAV=40. Again, the comparison run has removed the stochastic variability for graphical clarity, which explains the earlier winter bloom (see Fig. 54). The point of interest is that the severe surface inhibition ($I/I_{opt}=4$) that resulted during bright summer days had a relatively small effect in depressing production. This comparison is in marked contrast to the previous case (Fig. 55), where a constant *high* I_{opt} resulted in severe light limitation ($I/I_{opt}<1.0$) during the winter months.

10.2 The "Most Limiting" Nutrient

Within the phytoplankton compartment, the model determines the most limiting nutrient in the Leibig sense, based on the kinetic half-saturation constants for the species-group. To the extent that three nutrients are modeled which may potentially be limiting at any given time or place, multiple nutrient effects have been considered. However, true multiple interactions where synergism may alter conclusions based on a single nutrient concentration are poorly understood, and thus not represented in the model. An alternative to the scheme used here which attempts to reflect such effects, is the multiplicative interaction of the Monod

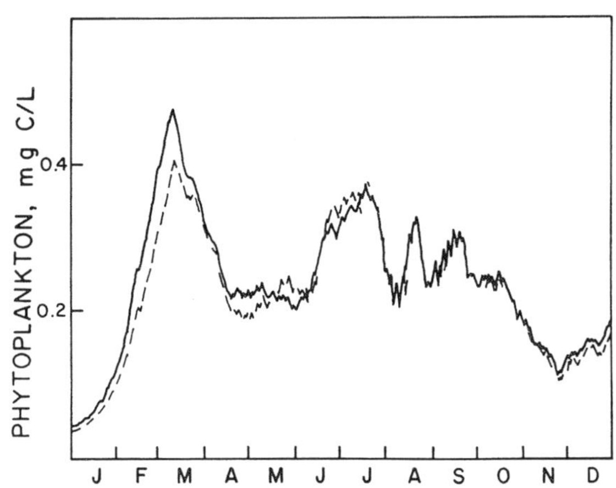

Fig. 57. Consequences of using a multiplicative nutrient limitation term, Eq. (9) *(dashed line)* rather than the single "most limiting" assumption of the standard run *(solid line)*

kinetic limitation fractions. The routine use of this formulation was rejected on theoretical grounds (Sect. 4.2), but the consequences of the assumption are of interest.

The anticipated effect in the model was a small but systematic lowering of the phytoplankton and zooplankton. However, the simulation was surprisingly similar to the standard run (Fig. 57). The detailed output revealed that the reduced growth in the winter–spring bloom proportionately raised nutrient levels compensating for the increased limitation in the summer. Again, the complexity of the model resulted in increased stability to minor alterations. Interestingly, another recent modeling study has concluded that the single versus multiplicative formulation made little difference, although the latter was used routinely (Lehman et al., 1975).

10.3 Two Species Groups and Luxury Nutrient Kinetics

Throughout this discussion the emphasis has been on the behavior of the simple one species-group phytoplankton compartment. The model also has the capacity to represent two groups, each with differing temperature and/or nutrient response characteristics. Two examples demonstrate the usefulness of this feature.

In a simple case, the species-groups were defined to represent summer and winter populations. Group one was assumed to be the same as the single population of the standard run. The second group was given a relative advantage in the summer by assigning a higher maximum growth rate than the Eppley curve at temperatures exceeding the bay average of 11.5° C. At colder temperatures, the growth was lower than the standard (Fig. 15A). In addition, the summer population was assumed to have a slight advantage with respect to nutrient kinetics, having a lower C:Si ratio and lower half-saturation constants for nitrogen and silica. As expected, the model demonstrates a clear seasonality with the standard group being clearly outcompeted during the summer months (Fig. 58). While the simulation was designed to demonstrate this changeover, it should be emphasized that the specified parameters for the first group were not altered in any way to make that population less productive than in the standard run. The dramatic succession produced in the model resulted only from the slight advantages conferred on the second group.

A more interesting simulation evaluates the significance of a hypothetical luxury uptake scheme. The potential importance of intracellular nutrient pools to the competitive strength of a phytoplankton species is a current topic of active research (Droop, 1973; Grenney et al., 1973). Detailed mechanistic hypotheses are being formulated which are ideally suited to simulation analysis. A recent paper investigated the dynamics of an elaborate physiological model of fresh-water algal species in substantial detail (Lehman et al., 1975). The Narragansett Bay model has been developed with flexibility in the phytoplankton nutrient kinetics to facilitate its use in such analyses.

Three assumptions about the nature of luxury kinetics were the basis for the hypothetical scheme used in the model (1) in highambient concentrations, cells will take up nutrients in excess of their minimum requirements, (2) the presence of these nutrient stores will enhance the growth rate at that time, (3) these cellular nutrients will allow growth to continue for a time beyond the point when ambient nutrients

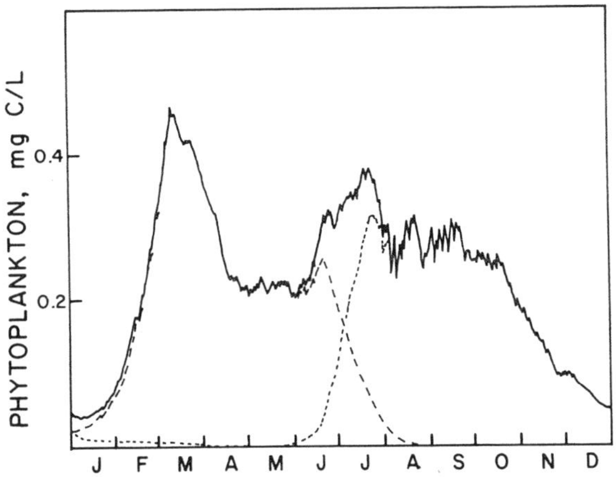

Fig. 58. Simulation of two species-groups, one with growth responses favored during summer conditions. A population favored by increased growth potential above 11.5° C (see Fig. 15 A) and more efficient nitrogen and silica kinetics *(dotted line)* outcompeted a population with characteristics of the standard run *(dashed)* during the summer. *Solid line* represents total phytoplankton and differs only slightly from the standard run

become depleted. An arbitrary scheme was constructed upon these premises which permits the uptake ratio of carbon fixed to nutrient assimilated to vary within a specified range, according to the value of the Monod limitation expression [Eq. (10)]. For example, when the ambient concentration equals the half-saturation value, the uptake ratio is half-way between specified extremes. In addition to this, the limitation term for the growth estimate (NLIM, PLIM) is determined as a function of any accumulated luxury pools in addition to the ambient levels. This composite value, NTOT1 in the program algorithm, acts to subsidize growth when luxury storages are great. The estimate based on the Monod calculation with NTOT1 determines the carbon growth, but the uptake, which is governed only by the ambient concentration, may or may not provide sufficient nutrients to maintain the luxury reserves. As ambient levels fall, enhanced growth continues at the expense of these reserves until they are depleted. When ambient levels remain low, growth-subsidy and uptake ratios are at a minimum, and the population will respond like the original, unmodified phytoplankton of the model.

The implications of this scheme for competition and for the seasonal patterns of abundance were interesting (Fig. 59). The two species-groups in the example differed only by the addition of luxury nitrogen kinetics to one; half-saturation constants, phosphorus and silica ratios, and temperature responses were identical. When ambient nutrients were high during the winter bloom, both species groups grew equally well, having each started with half the biomass. As ambient nutrients were depleted in the spring, the competitive advantage became evident. In addition, the luxury reserves in combination with the capacity for growth requiring less nitrogen resulted in a pronounced bloom in the early summer. While this scheme is both simple and hypothetical, it indicates the potential importance of including realistic biological detail, as well as demonstrating the ability of the model to evaluate such features.

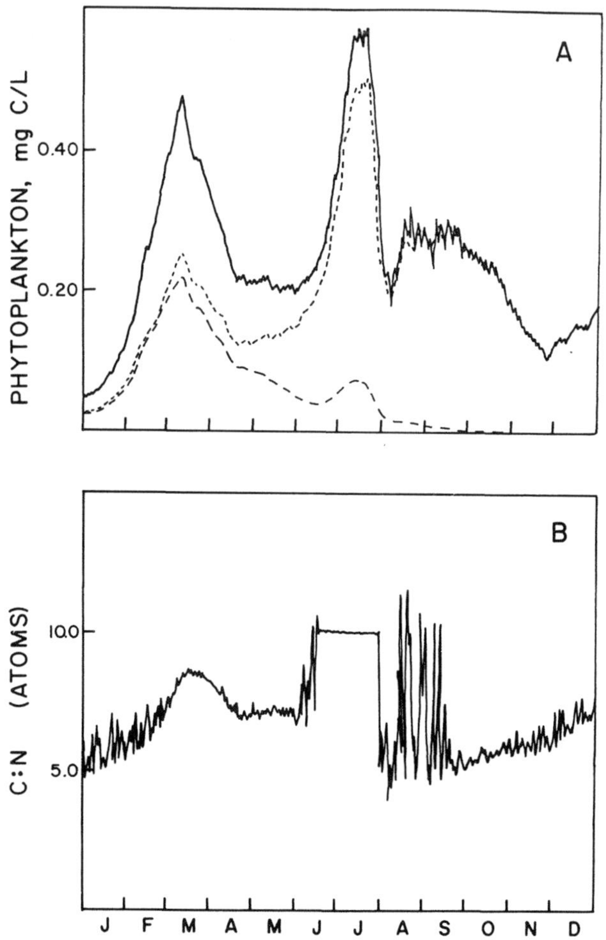

Fig. 59 A and B. Luxury uptake example. Two species groups were simulated with equal attributes except for the addition of a luxury nitrogen uptake capacity. (A) Starting with equal biomass, the competitive advantage of the intracellular storage *(dotted line)* became significant as nutrients were depleted in the spring. The stored nitrogen and the capacity to grow with a higher C:N ratio permitted a second bloom in the early summer despite low ambient nutrients. (B) The C:N ratio of the species group with the luxury uptake capability varied around the constant value of 7.0 of the other. The constant high ratio occurred during the summer bloom when nitrogen was being used most efficiently

10.4 Juvenile Zooplankton and Cannibalism

The major innovations within the zooplankton compartment of this model were the inclusion of detailed population structure and the omnivorous trophic role of the copepods. The significance of particulate carbon food subsidy in maintaining low winter populations has already been discussed.

The role of the time lags resulting from the separation of adult reproduction, egg hatching, and juvenile development was dramatic. When the standard run was

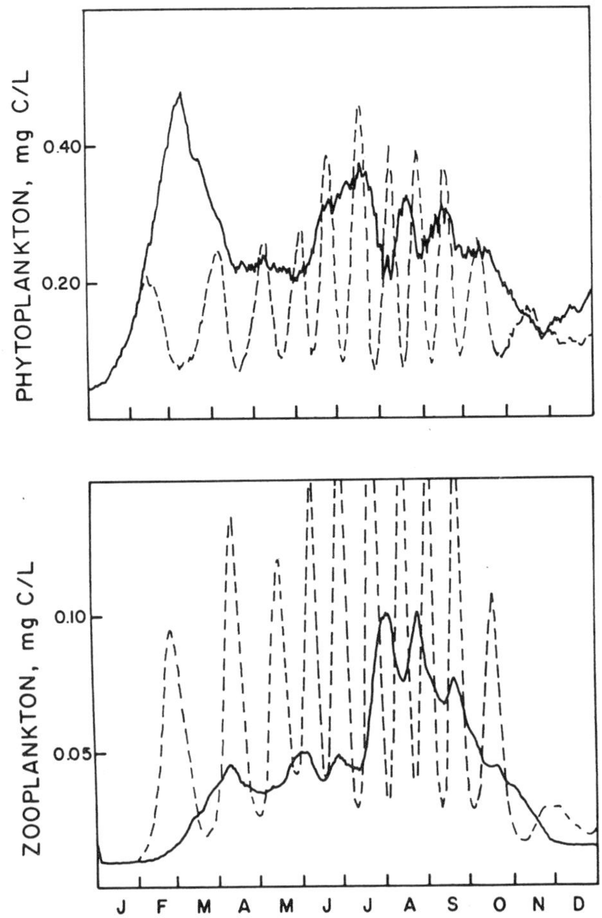

Fig. 60. Simulation with a single, homogeneous zooplankton compartment. Any unrespired assimilation was added directly to the adult biomass with no time lag *(dashed line)*. The regulatory role of egg and juvenile time delays in the standard run *(solid line)* is very strong

modified to a single composite zooplankton compartment, so that adult unrespired assimilation was immediately recruited as increased adult biomass, the model developed immediate, extremely rapid cycles which persisted throughout the year (Fig. 60). As soon as the winter bloom began, zooplankton responded rapidly, terminating further algal growth and initiating undamped but stable oscillations. Reducing the daily ration tended to lessen this effect, and sufficiently low daily ingestion stabilized the model noticably. With the delayed coupling due to generation times, the model is less erratic, even with adult rations approaching as much as 1.6 mg C/mg C daily at 20° C (RMX0=0.5, Q10RMX=1.8; see Sect. 11.1.5). The significance of delays in predation or grazing cycles is well known, and these model results are not new. They do suggest, however, that models without realistic time delays may achieve stability by underestimating herbivore ingestion.

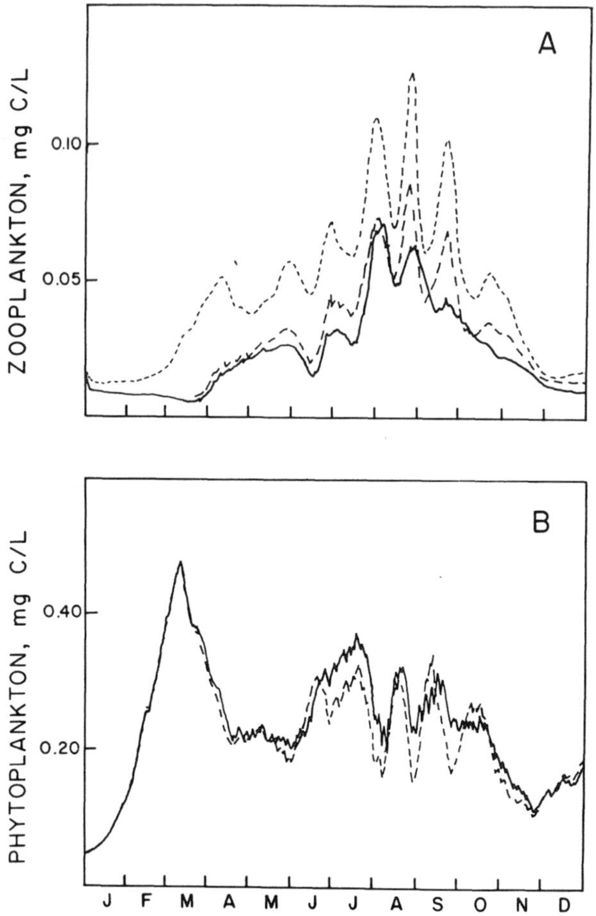

Fig. 61 A and B. The effect of removing cannibalism by adult zooplankton on juveniles from the model. (A) The total zooplankton *(dotted line)* and adults plus juveniles more than 50% mature *(dashed line)* for the no cannibalism run were compared to the adults plus mature juveniles of the standard run *(solid line)*. (B) The effect of cannibalism in stabilizing the phytoplankton of the standard run *(solid line)*

Although the potential carnivorous nature of smaller copepods has been recognized for some time, models have usually characterized them as simple herbivores, with the assumption that this trophic role was dominant. This model investigated this assumption by assuming that adult zooplankton could ingest eggs and juveniles up to a certain developmental stage. This formulation stabilized the phytoplankton–zooplankton cycles during the summer, in addition to generally lowering the adult biomass throughout the year (Fig. 61). The total zooplankton biomass in a simulation excluding cannibalism of juveniles entirely (upper curve, dotted line) showed an increased frequency and amplitude of pulses. In addition, both the total and adult-component biomass demonstrated more extreme oscillations. These features are also reflected in the phytoplankton, where accelerated early zooplankton production increased grazing during the first summer bloom, and late summer cycles occur with increased amplitude (Fig. 61B). The stabilizing effect develops because juveniles recently produced when phytoplankton are decreasing do not continue to develop. Rather, adult predation removes them from the system, thus simultaneously lowering grazing pressure and augmenting adult metabolic needs.

10.5 The Role of Carnivores

A succession of pulses of carnivores exert predation pressure on the zooplankton in Narragansett Bay (see Figs. 30–32). In the model, the seasonal influences of fish larvae, ctenophores and menhaden have been included to assess their role in the plankton system. As might be anticipated from the previous discussion of cannibalism, this carnivorous feeding exerts a significant controlling influence on the general stability of the total system. The biomass estimates (Chap. 6) show that this carnivorous pressure is important in the bay during the summer months when the model indicates a potentially unstable condition. It is an effective strategy to tap the high-turnover system which can rapidly respond to the losses, but the effect also confers more general stability on the entire community.

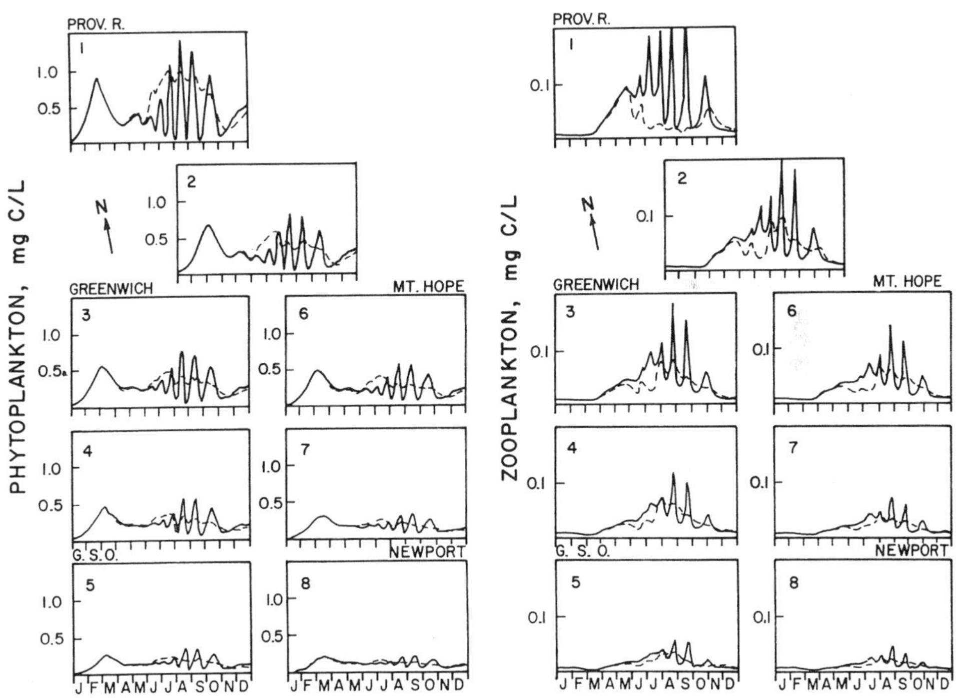

Fig. 62. Role of carnivores in stabilizing phytoplankton-zooplankton patterns in the model. Removing all carnivores substantially enhances algae-herbivore cycles in the summer in comparison to the standard run *(dashed line)*

When all carnivores, excluding adult cannibalism, were removed from the model, dramatic departures from the standard run were observed (Fig. 62). Fish larvae are the first to enter the system, accounting for additional predation early in the summer. Ctenophores and menhaden together exert substantial pressure throughout the remainder of the summer and fall. A comparison run removing only the menhaden indicated that their influence was dominant, especially in the upper bay, and their stabilizing effect was evident even at the bay mouth. Ctenophore predation also significantly depressed late summer fluc-

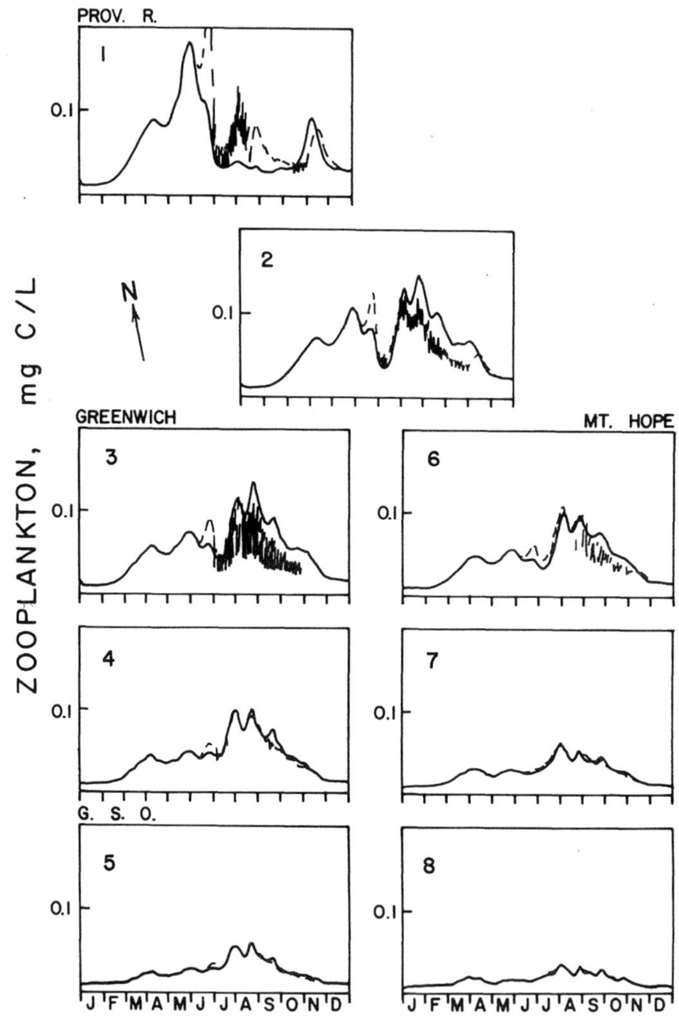

Fig. 63. The simulated impact on total zooplankton when schools of menhaden were assumed to seek out the region of highest zooplankton concentration in which to feed. The "hunters" moved rapidly among the upper bay elements/from July through October, severely depleting zooplankton levels in each day's target element. Mixing from adjacent areas on subsequent days was almost sufficient to replenish the stock

tuations, however, even in the Providence River, the region of greatest menhaden impact.

The role of carnivores has rarely been included in mechanistic models, especially not in models with detailed algal-herbivore dynamics. Further, the uncertainty of the menhaden biomass data made it interesting to investigate other possible baywide patterns. Numerous variations have been simulated, with changes in the abundance, seasonal timing, and spatial distribution of the menhaden. In general, these results were not surprising. For example, increased or decreased abundance of the fish predictably allowed relatively or lower higher zooplankton stocks to develop, along with the previously mentioned tendency for lesser or greater oscillations. Altering the seasonal pattern of abundance similarly shifted the period of depressed zooplankton, although the extent of such depressions could be markedly different when, for example, menhaden predation was superimposed on the peak fish larvae abundance. One series of runs proved to be especially interesting. To avoid the artificiality of forcing smooth curves of abundance, runs were made in which concentrations of the carnivores were moved around the bay simulating intensive predation pressure by migrating schools of menhaden. When a given total biomass was moved among any number of elements on a regular schedule, each region was severely affected, with pronounced spikes occurring sequentially in the zooplankton patterns. In contrast, then, intensive predation is clearly disruptive when applied intermittently to separate patches within an area.

In the last run of this series, 15×10^6 pounds of menhaden, the same total biomass as the standard run, were allowed to migrate freely throughout the bay from July through October, seeking the highest food concentration. Each day, the entire fish population was assumed to feed in that element of the bay with the greatest total zooplankton. Again, the resulting zooplankton pattern showed dramatic irregularities (Fig. 63). While the menhaden remained exclusively in the upper half of the bay, a detailed look at the frequency of their appearance in the bay reveals an interesting feature.

Percent of time spent by Menhaden in each spatial element

Month	1	2	3	4	5	6	7	8
July	35.5	25.8	25.8	9.7	—	3.2	—	—
August	16.1	51.6	25.8	—	—	6.5	—	—
September	—	43.4	33.3	3.3	—	20.0	—	—
October	16.1	35.5	25.8	3.2	—	19.4	—	—
Overall	17.0	39.0	27.5	4.5	—	12.0	—	—

During July, the fish concentrated their feeding in the Providence River, spending less time in elements 2 and 3. This agrees well with the observations of commercial fishermen, which we used to assign the distribution for the standard run. In August and September, however, simulated menhaden pressure switches sharply away from element 1, as can also be seen by careful inspection of Figure (63). This is because of heavy competition by the large population of ctenophores which effectively excluded the fish from the most productive element.

While the formulations for menhaden and the other carnivores in the model are not elaborate, the importance of higher levels in altering seasonal patterns and in reducing instability is clearly indicated. There is, of course, a chance that the forced nature of these compartments enhances this result, and it would be interesting to determine whether mechanistic formulations of appropriate feedback influences for these compartments would retain this general character. Menhaden would be inappropriate for such feedback dynamics, since most of their life is spent outside the bay. But a mechanistic model of ctenophore population dynamics has been developed for the ctenophore, *Mnemiopsis leidyi*, in Narragansett Bay (P. Kremer, 1976) which uses forced zooplankton levels. Future modifications of both programs may couple these models and allow a more complete analysis of carnivore effects without relying on forced inputs of either predators or prey.

11. Sensitivity and Stability

11.1 Sensitivity Analysis

Of the many coefficients and parameters used to specify rates and interactions in the Narragansett Bay model, few, if any, may be assigned a numerical value with a high degree of certainty. While Narragansett Bay is one of the most intensively studied estuaries in the world, numerous critical details about the bay system and the functioning of its components are not yet available. Thus, it was often necessary to draw on literature from more or less distantly related systems. Unfortunately, there is a wide range in many of the measurements that have been reported (e.g., see Table 8). This uncertainty is not unexpected if one considers the large amount of biological, spatial, and temporal variability that is characteristic of natural systems. In addition, there are numerous technical difficulties that one encounters when trying to assess biological processes in the laboratory or in the field. For these reasons, the choice of any one value for a coefficient may be questioned. While we have tried to provide reasonable justification for the values used in the standard run of the Narragansett Bay model, we were also curious to find out how sensitive the simulation was to those particular choices. The process of varying coefficients, forcing functions, initial conditions or certain aspects of the computer program not only provides insight into the behavior of the model, but into the more complex natural system as well. It also serves as a crucial feedback loop in suggesting sensitive areas where additional research is needed. In theory, it may also discourage the collection of redundant or peripheral data.

11.1.1 Initial Conditions

One of the first sensitivity runs was made to determine the importance of the initial conditions to the computed seasonal patterns of the model. Since the field data used as input on day 1 of the simulation were considered accurate to within a factor of 2 in representing the average conditions in each element, runs were made with one half and double the initial conditions used in the standard run. The results showed that these changes did not alter the timing or characteristics of the seasonal pattern (Fig. 64); only the magnitude of the winter–spring bloom and associated zooplankton crop was affected.

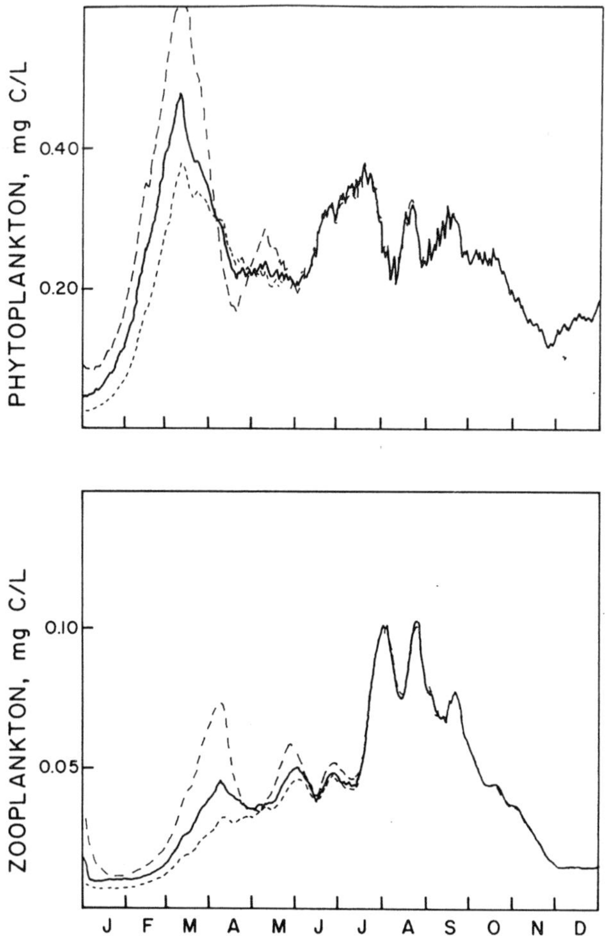

Fig. 64. Effect of increasing and decreasing the initial conditions of the standard run by a factor of two. Only element 4 is shown, as similar changes occurred throughout the bay

11.1.2 Hydrodynamic Characteristics

The inclusion of some degree of spatial heterogeneity and hydrodynamic complexity in the Narragansett Bay model appeared to have been successful in reproducing the major distribution gradients observed in the bay. However, it was not clear from the standard run how important these features were to the stability of the biological interactions of the model. Simulations run without any physical mixing showed pronounced phytoplankton–zooplankton cycles as well as dramatic changes in the timing and magnitude of the winter–spring bloom (Fig. 65). Without the influence of flushing, there was no decline in the standing crop down the West Passage, though the much greater depth and volume of the East Passage continued to result in lower concentrations of materials. The importance of depth–light interactions in initiating the winter–spring phytoplankton bloom is much clearer here than in the standard run. Without tidal mixing, the bloom might appear in the

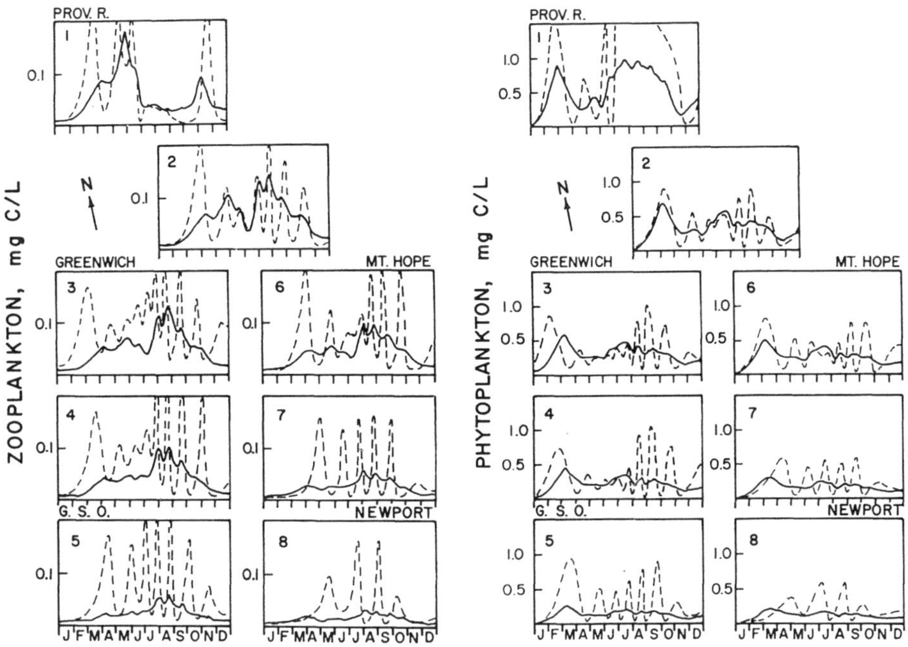

Fig. 65. Effect of removing all flushing and water exchange *(broken line)* from the standard run *(solid line)*. The eight spatial elements were simulated as individual isolated systems

upper West Passage three to four weeks ahead of the lower West Passage and two months ahead of the lower East Passage.

As discussed in Chapter 3, it is very difficult to make direct measurements of the net exchange rate between Narragansett Bay and Rhode Island Sound. At that point, we went through various indirect arguments to arrive at a range of exchange estimates. In order to get a feeling for the possible implications of this uncertainty, simulation runs were made using the maximum and minimum estimates. While the results indicate that neither the phytoplankton dynamics nor the nutrient cycles are particularly sensitive to changes in flushing rate of this magnitude, some interesting contrasts in zooplankton were obtained (Fig. 66). When the minimum estimates of the exchange factors were used, the increased retention in the bay advanced the onset of the grazing cycles and increased their amplitude. This effect was especially apparent in the lower bay, where a significant amount of the spring zooplankton variation was lost from the simulation.

11.1.3 Extinction Coefficient

The nonchlorophyll-related portion of the extinction coefficient (k_0, see Sect. 4.3.4) is another physical characteristic of the bay that is difficult to describe with any one value. Measurements by Schenck and Davis (1972) have shown that the spatial variation in k, and presumably k_0, is complicated by changes during the tidal cycle as sediments are resuspended. While the model includes the

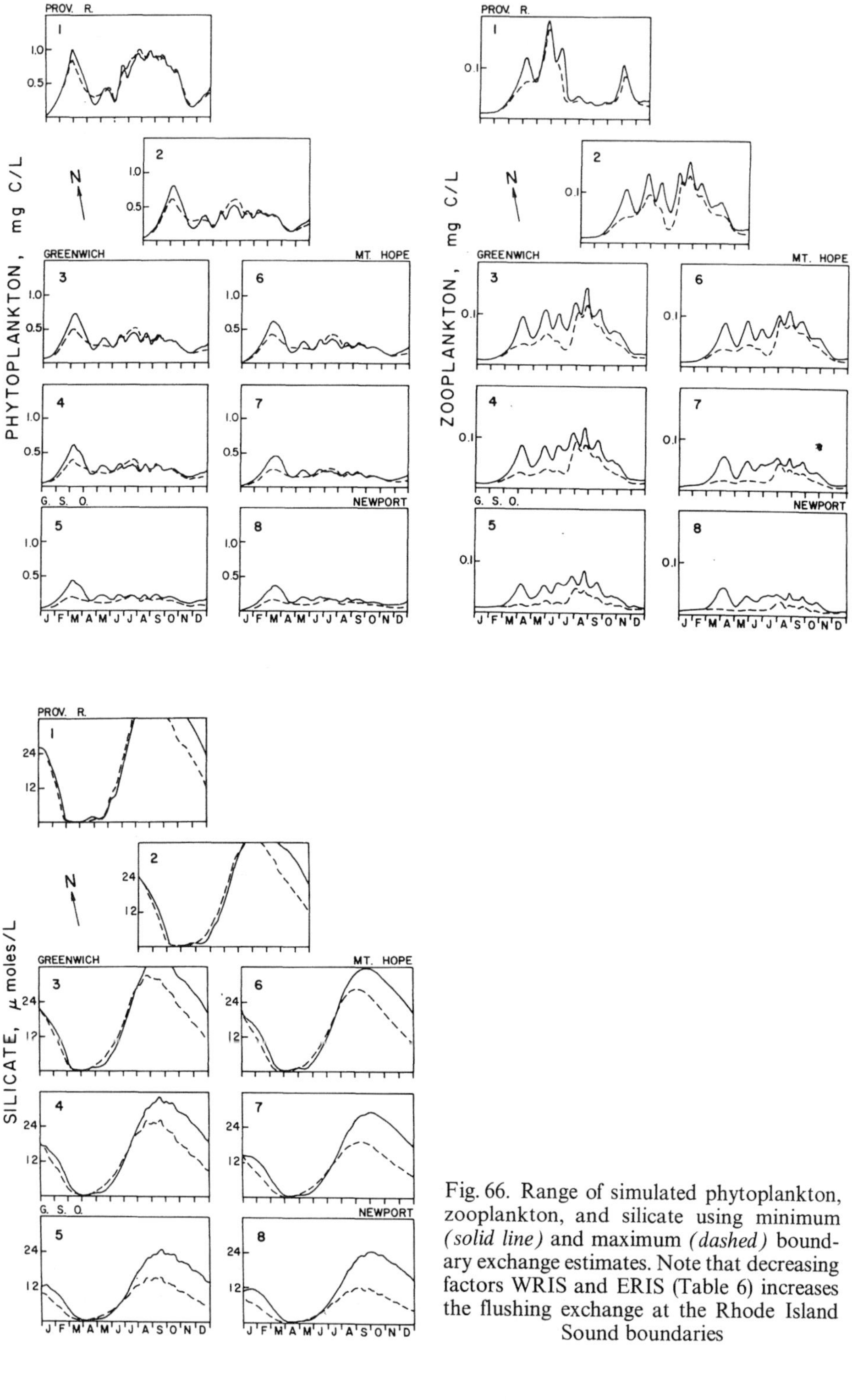

Fig. 66. Range of simulated phytoplankton, zooplankton, and silicate using minimum *(solid line)* and maximum *(dashed)* boundary exchange estimates. Note that decreasing factors WRIS and ERIS (Table 6) increases the flushing exchange at the Rhode Island Sound boundaries

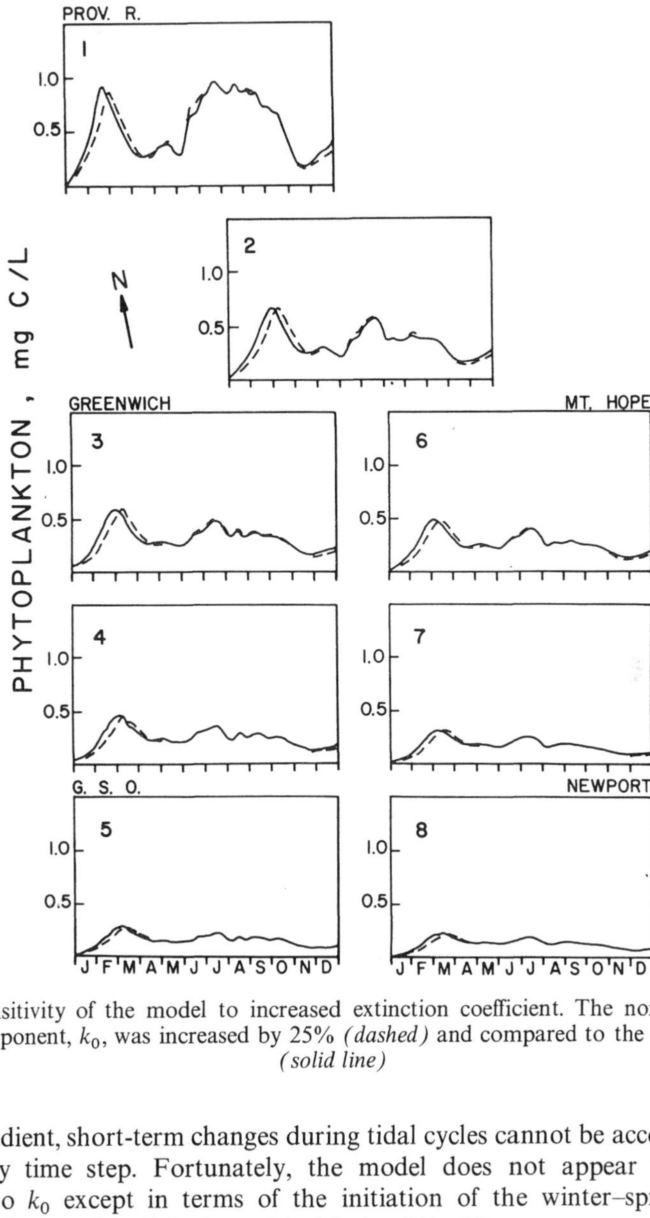

Fig. 67. Sensitivity of the model to increased extinction coefficient. The nonchlorophyll-related component, k_0, was increased by 25% *(dashed)* and compared to the standard run *(solid line)*

spatial gradient, short-term changes during tidal cycles cannot be accommodated in a 1-day time step. Fortunately, the model does not appear to be very sensitive to k_0 except in terms of the initiation of the winter–spring phytoplankton bloom, when light is a critical factor (Fig. 67).

11.1.4 Phytoplankton

A number of the basic biological processes in the model required sensitivity analysis. While the maximum growth rate of the phytoplankton in the model is determined as an exponential function of temperature using a formulation given by

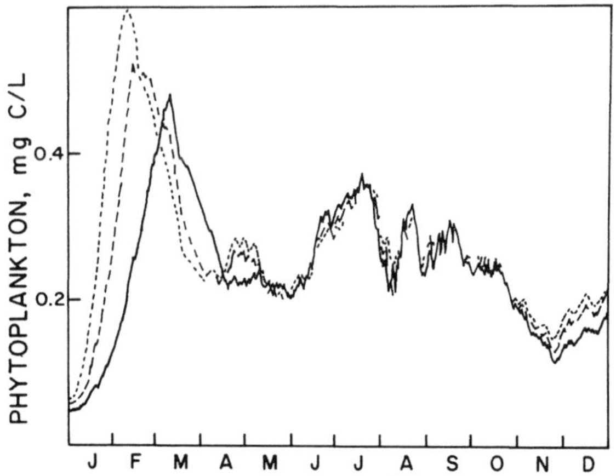

Fig. 68. Effect of increasing the temperature-dependent maximum growth rate of the phytoplankton by 0.25 *(broken line)* and 0.50 *(dotted line)* divisions per day compared to the standard run *(solid line)*. Only element 4 is shown

Eppley (1972), a number of measurements of growth rates exceeding this formulation have been obtained. If the value of the Eppley curve is increased by 0.25 or even by 0.5 division per day, however, the effect on the model simulation is not substantive. The only impact of higher maximum growth rates of the phytoplankton appears to be an increase in the initial rate and intensity of the winter–spring bloom (Fig. 68).

In an earlier section (Sect. 10.3) the effect of varied nutrient:carbon ratios in adjusting the summer standing stock of phytoplankton was discussed in some detail. Considering this critical role of nutrients, and of nitrogen in particular (Fig. 47), an obvious parameter to vary in a sensitivity analysis was the kinetic nutrient

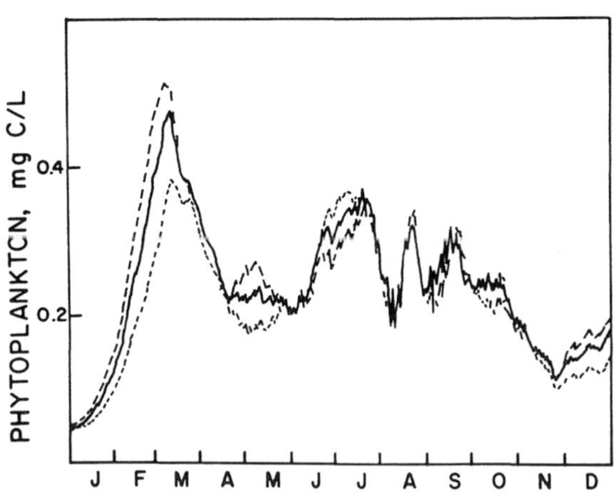

Fig. 69. Sensitivity of the simulated phytoplankton to changes in the nitrogen half-saturation constant. Standard run $K_s = 1.5$ µg-at N/l *(solid line)*, $K_s = 0.75$ *(dashed)*, $K_s = 3.0$ *(finely dashed)*. Only element 4 is shown

parameter, K_s. Figure 69 compares simulations with half and double the value of the nitrogen half-saturation constant. The results are not surprising. Even when nutrients were relatively available during the winter, the Monod formulation [Eq. (10)] conveyed an advantage to the lower K_s value, in this case the heavily dashed line for $K_s = 0.75$ µg-at N/l. Interestingly, the high value, $K_s = 3.0$, consistently gave the lowest simulation except during a period in July, when it exceeded the predictions of the other two runs. This apparent anomaly developed from low spring production and the associated reduction in zooplankton biomass. The phasing of the herbivore increase and the onset of the summer grazing cycles resulted in sufficiently relaxed pressure on the phytoplankton to allow a brief flowering. This sort of detailed analysis of the simulations should not be interpreted as giving definitive predictions of a specific behavior in reality. Clearly the uncertainties are too great

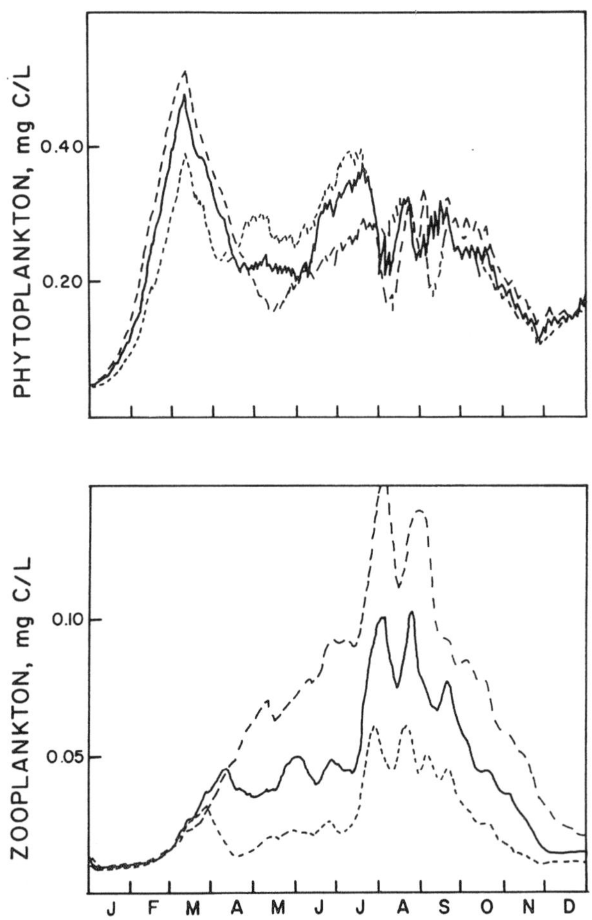

Fig. 70. Sensitivity of the simulated phytoplankton and total zooplankton to simultaneous changes in both adult maximum ration and respiration. Standard run *(solid line)* vs. half *(dashed)* and double *(dotted)* the parameters RMX0 and XRSP0. Only element 4 is shown

to attach precise significance to detailed comparisons. Rather, such analyses may provide insights into the sorts of processes that can emerge from the complexity of feedback controls in natural systems. While these insights are rarely entirely counterintuitive, they are often unexpected, and the possibilities suggested by simulations should be evaluated as potential factors in explaining natural phenomena.

11.1.5 Zooplankton

Within the zooplankton compartment, two parameters which are of basic importance are the maximum ration and the respiration of the adults. These temperature-dependent rates initially control secondary production in the model. The phytoplankton and zooplankton patterns simulated with one-half and two times the values of both adult zooplankton ration and respiration used in the standard run are presented in Figure 70. It is encouraging that while significant differences in the magnitudes result, the seasonal patterns are quite similar. Note that decreasing both parameters simultaneously increases the turnover time of the zooplankton compartment and decreases the grazing pressure. As a result of the latter, phytoplankton standing crop is increased. The higher food supply and slower turnover time act together to increase zooplankton biomass through most of the year.

It is also interesting to vary only one of the two parameters discussed above, so that the ratio of zooplankton respiration : maximum ration varies. The results of changing this efficiency ratio 25% by altering the maximum ration or respiration independently indicate that the model is more sensitive to the ratio of the parameters than it is to the values of the parameters per se (Fig. 71).

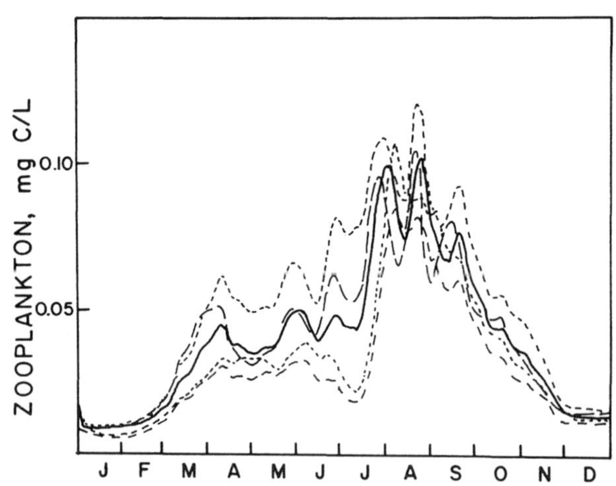

Fig. 71. Effect on the simulated zooplankton standing crop of varying the ratio of adult maximum ration to respiration by ±25%. The higher population levels resulted from increasing the ration or decreasing the respiration, while the lower values were produced by opposite changes. The standard run in element 4 is shown with a *solid line* for comparison

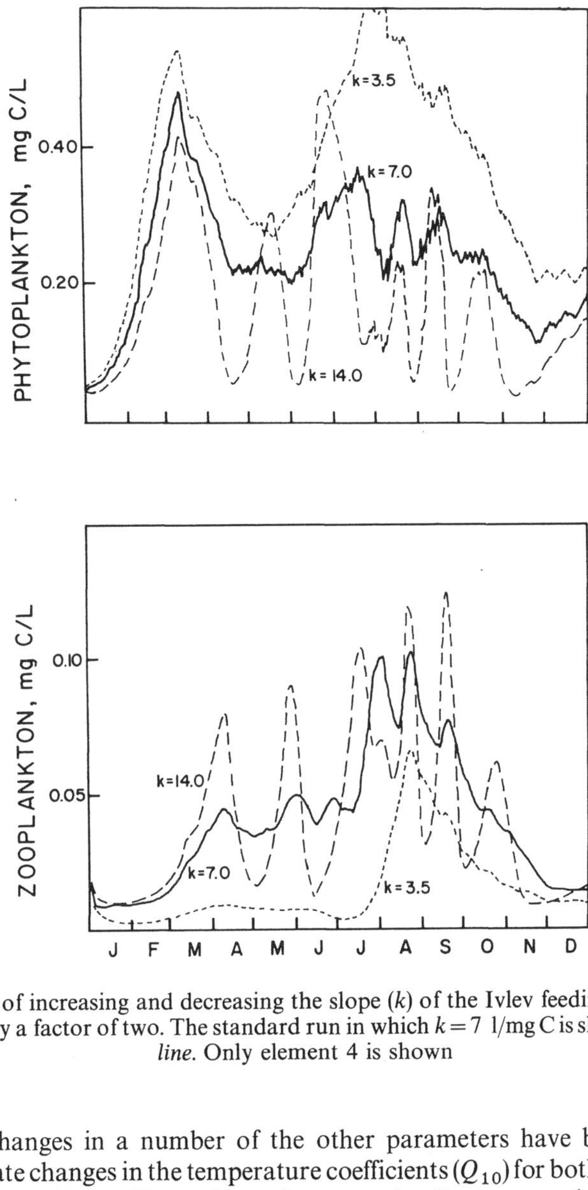

Fig. 72. Effect of increasing and decreasing the slope (k) of the Ivlev feeding hyperbola for zooplankton by a factor of two. The standard run in which $k=7$ 1/mg C is shown with a *solid line*. Only element 4 is shown

Similar changes in a number of the other parameters have been evaluated. Commensurate changes in the temperature coefficients (Q_{10}) for both ingestion and respiration have little effect, although unequal adjustments alter the relative reproductive efficiency in warm and cold seasons. Higher Ivlev food-limitation exponents (k, Fig. 28) increase the amplitude of the phytoplankton–herbivore cycles by allowing more effective grazing to occur, thus lowering phytoplankton concentrations (Fig. 72). Lower Ivlev terms reduce the ration attainable by the animals and may even result in extinction. Changes in the juvenile maximum growth rate and development time may significantly alter the time lags which govern the initiation and severity of grazing cycles. Results of the extreme case, in which the time lags were reduced to zero, have been discussed earlier (Fig. 60).

11.1.6 Seasonal Cycles

At this point, all of the runs discussed as part of the sensitivity analysis have been concerned with varying individual coefficients or processes. It is also interesting to examine the response of the model to changes in the basic forcing functions of temperature and light. If the seasonal temperature cycle is removed, and the standard run is executed with a constant temperature equal to the annual mean (11.5° C), only a few of the major patterns of the seasonal phytoplankton cycle remain, and the basic zooplankton cycle is lost entirely (Fig. 73). On the other hand, if runs are made with the seasonal temperature cycle present and the light input held constant, the results are quite different. As the constant light input is increased from a very low 25 ly per day, the simulations increasingly come to resemble the standard

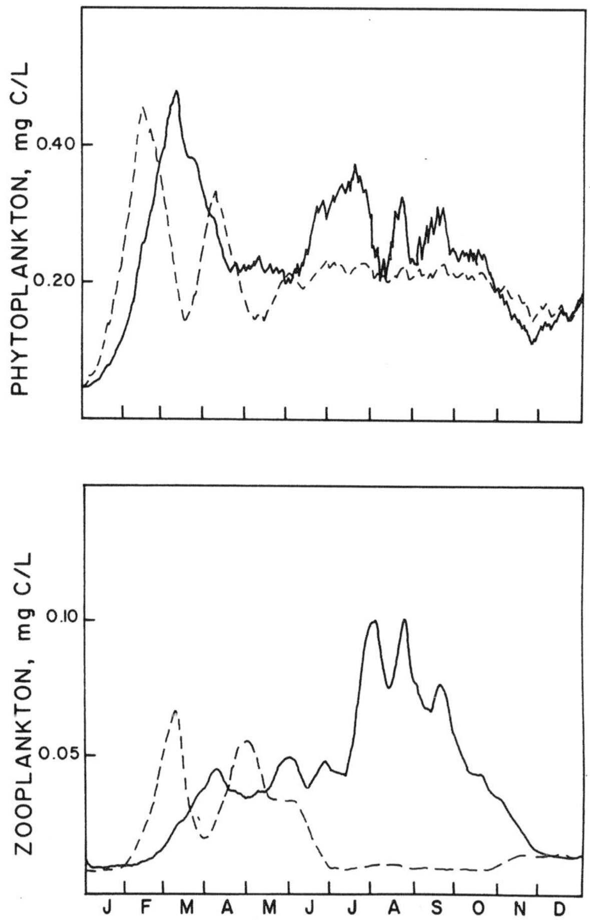

Fig. 73. Effect on simulated phytoplankton and zooplankton populations of removing the seasonal temperature cycle *(broken line)* from the standard run *(solid line)*. While all other forcing functions varied as in the standard run, temperature was held constant at the annual mean of 11.5° C. Only element 4 is shown

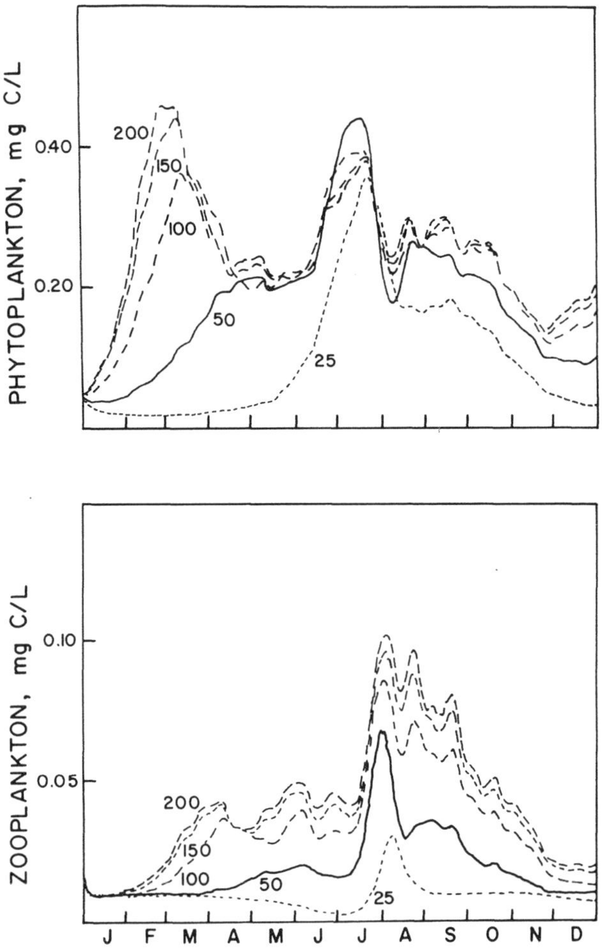

Fig. 74. Effect on simulated phytoplankton and zooplankton populations of removing the seasonal cycle of light input from the standard run. All other forcing functions varied as in the standard run, but light was held constant at 25, 50, 100, 150, and 200 ly/day. Only element 4 is shown

run (Fig. 74). When the constant daily input is between 200–300 ly, the seasonal cycles of both phytoplankton and zooplankton are very similar to those of the standard run (Fig. 75). The most obvious trend in Figure 74 is the progressive increase in magnitude and earlier occurrence of the winter–spring bloom as the light level increases. The importance of light in the initiation of the bloom has also been shown in field and laboratory studies by Smayda and Hitchcock (1977) and by Nixon et al. (in press). It is also clear from Figure 74 that the light saturation feature inherent in the basic physiological response emerges as an important part of the unified behavior of the system, even after time and depth integration of photosynthesis.

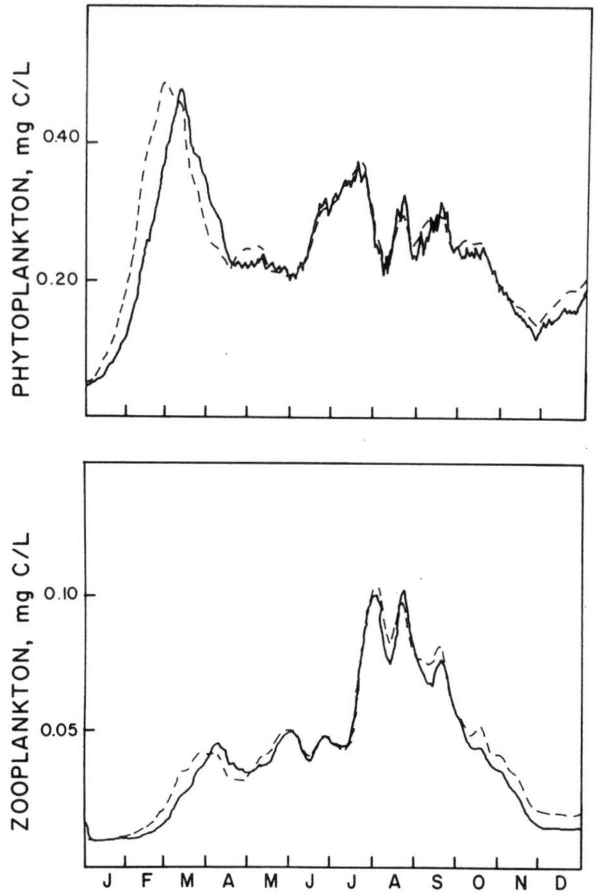

Fig. 75. Comparison of the standard run *(solid line)* with the results of a simulation in which light input was held constant at 300 ly/day, the annual mean *(broken line)*. Only element 4 is shown

11.1.7 General Implications and the Role of Sensitivity Analysis

In general, the variety of responses which the model produces with reasonable parameter choices is not extreme. The initial phytoplankton bloom and the summer oscillations are general features, and varied parameter combinations usually alter only the frequency, relative magnitude, and seasonal duration. As has been stated, perhaps the most interesting results are the inabilities of the model to produce certain observed patterns, including the fall phytoplankton decline and the stable, low zooplankton population during the winter. In such cases, the model is strongly indicating that these patterns are not explicable solely with the factors presently included, and additional investigation into these areas seems desirable.

Sensitivity analyses reflect the properties of the model, but they may also have implications for the more complex natural system. Moreover, the direct comparison of simulations with observed data, and the determination of which parameter choices improve the agreement is also a useful tool in ecosystem analysis. A frequent objection to this type of model analysis argues that the numerous parameters and equations result in a system response which is so malleable that any set of observations may be fitted. Such an argument is only valid for a strictly mathematical approach using equations with no foundation in ecological understanding. Then, in fact, n observational points may be fitted with an equation of $n-1$ coefficients. One approach even takes advantage of this property, postulating equations with rudimentary biological significance and interpreting the implications of the parameters required to achieve a statistically accurate fit (Ulanowictz et al., 1975). In mechanistic models such as this one, the forms of the equations and the relationships represented by their interactions are carefully designed to be ecologically meaningful. Further, the parameters are necessarily constrained within reasonable values based on natural or laboratory observations. In this system, agreement with a set of seasonal observations is by no means assured, although the possibility of spurious agreements resulting for the wrong reasons must be considered.

In fact, the addition of increased biological detail contributes in one sense to uncertainty, and at the same time enhances the reliability of the criteria for success. For example, a very simple model which is a poor representation of the system also provides very little information about the underlying processes resulting in a given prediction. Agreement with a simplified set of observations may be the only test of the modeler's assumptions. In a very detailed formulation, numerous internal rates, ratios, etc., are available for verification in addition to the pattern of the state variables. Now, not only are the emergent properties of the system available, but detailed information about the numerous components may also be critically examined. Should satisfactory agreement for all these detailed compartments and rates be attained, the probabilities of spurious agreements are substantially reduced. Precisely because the model represents an imperfect analog of the system, "complete agreement" is an unreasonable expectation. Rather, it is the inability of the model to reproduce certain observed behavior that may be the most interesting and informative result. If no combination of reasonable parameters results in some specific pattern, then either the model does not include, or incorrectly includes, the pertinent factors and interactions. An alternative explanation is that the observation was invalid. While this is not frequently offered as a serious justification of disparate model results, the uncertainty in the data base which a model is built upon and "tested" against is an underlying consideration of major importance. Modeling may even play a role here, suggesting with greater precision than can ever by achieved in the field the spectrum of behavior which may be expected from a reasonable range of measured processes.

A final note should be added concerning sensitivity evaluation of ecosystem models. As has been noted, the systematic varying of single parameters which normally constitutes "sensitivity analysis" is both cumbersome and difficult to interpret meaningfully in an ecological context. Yet the wide variability of virtually all reported values makes some consideration of this problem essential. Another traditional approach has been to incorporate a stochastic aspect into parameter choices using Monte Carlo simulation. In this approach, values are selected at random, or conforming to some statistical distribution, and one complete simulation is run. The average results of many such simulations are then interpreted as some measure of sensitivity to variability. This method has recently been employed with ecosystem models. A related alternative is presently being developed for use with our model. Instead of only varying parameter values at the start of each run, it may be more desirable to incorporate variation throughout the simulation. By specifying a coefficient of variation (standard deviation as a function of the mean) for each value, a Gaussian number generator can simultaneously vary any number of parameter values independently each day of the simulation. It is an interesting observation that natural systems appear remarkably stable despite wide, short-term variability in what may be only loosely coupled responses. Models, in general, are more sensitive to such changes, although this program suggests that increased complexity with ecological relevance may reduce this problem. The stochastic sensitivity scheme under development should provide some insight into the emergent stability of intrinsically variable systems.

11.1.8 A Note on Numerical Stability

The numerical stability of a detailed set of complex and highly coupled nonlinear differential equations such as those found in the Narragansett Bay Model

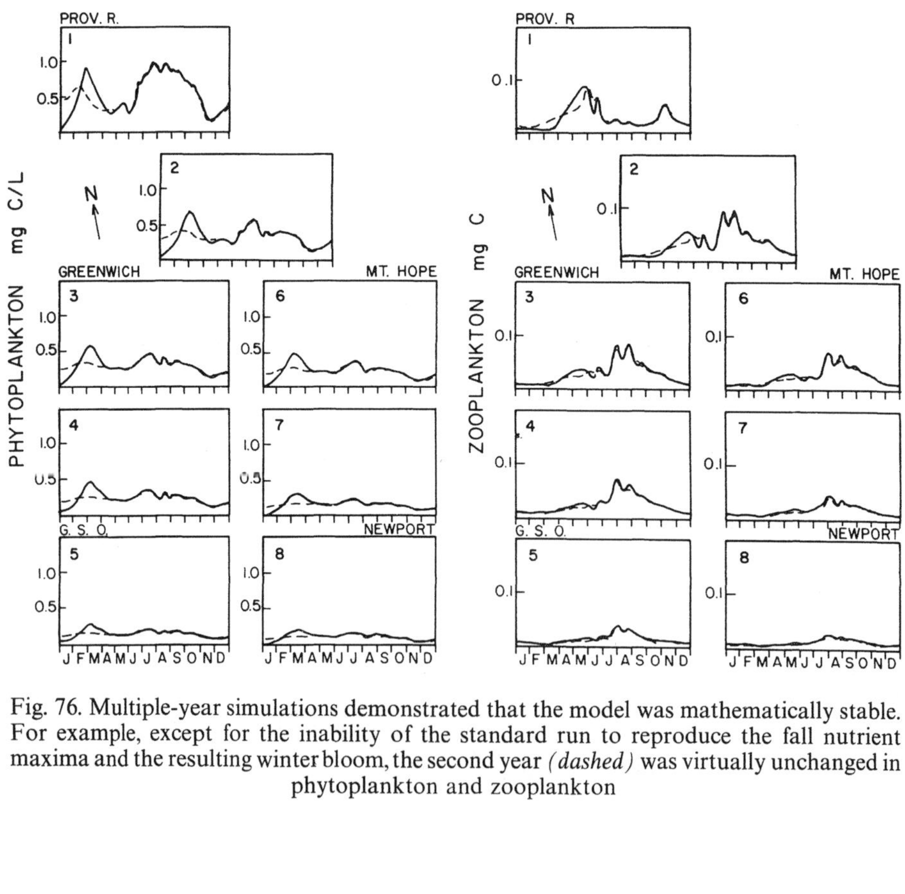

Fig. 76. Multiple-year simulations demonstrated that the model was mathematically stable. For example, except for the inability of the standard run to reproduce the fall nutrient maxima and the resulting winter bloom, the second year *(dashed)* was virtually unchanged in phytoplankton and zooplankton

is of mathematical interest. For the conditions of the standard run, multiple-year simulations were carried out, and successive patterns were virtually unchanged (Fig. 76). As discussed earlier, the fall simulation disagrees with reality, and the lack of a nutrient maximum precludes repetition of the winter phytoplankton bloom. As strictly a mathematical system, however, the timing and magnitude of the oscillations are quite reproducible, returning to the same conditions at the end of each simulated year. For sets of parameters that induce more extreme oscillations, the same assertions may be made, although there was a slight tendency for amplitudes to diminish in long simulations.

A number of different starting times have been tested in the model. Initially, runs were begun on August 1 to simulate the actual sampling year. Again, the fall

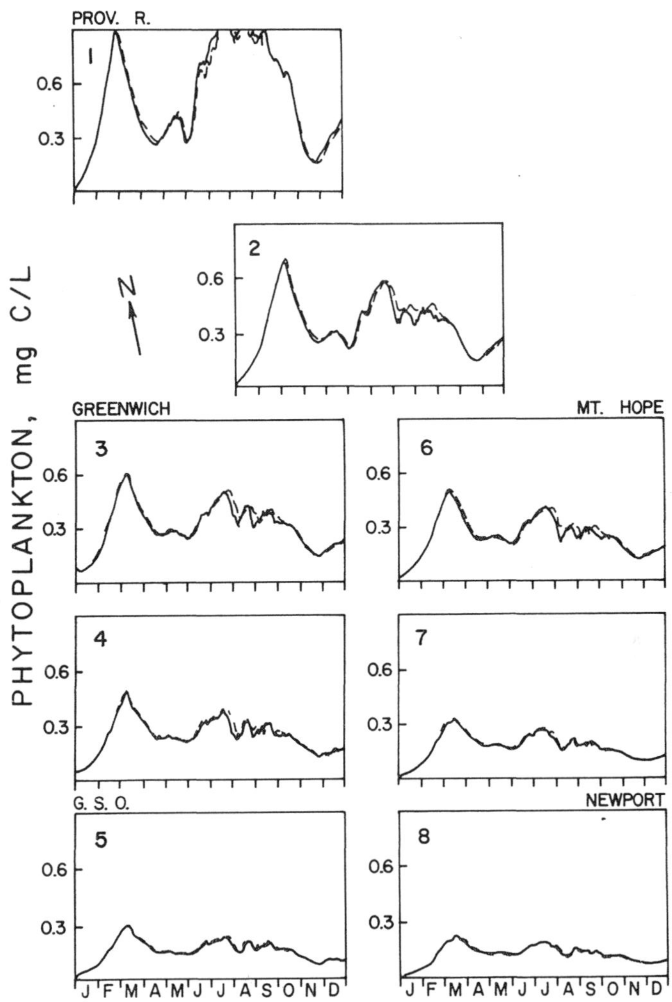

Fig. 77. Comparison of the standard run with the predictor–corrector integration *(solid line)* vs. a simpler one-step method

deficiencies of the nitrogen and phytoplankton compartments decreased the utility of these cases, but as expected, the simulations retained the general stability characteristics of the multi-year runs. Initiating the model at the peak of the winter–spring bloom when nutrients were already depleted also returned reasonable results. Following an initial zooplankton increase that was more rapid than in the standard run where food limitation was severe, the model adjusted to appropriate levels for all state variables.

The stability and consistency of the model is, in a basic sense, a reflection of the fact that the natural system approaches almost a limiting condition during a significant part of the year. The model's approximate representation of the spring season biomasses under almost complete nitrogen deficiency acts as a limit around which various simulations converge. While this feature is a mathematical characteristic in the model, it is an analog of the control exerted by the repeating seasonality of temperate ecosystems.

In view of the relatively long time step used in the model, it was also of interest to determine the sensitivity of the model to the method of solution. In anticipation of potential problems due to the one-day integration interval, a predictor–corrector scheme was employed (Sect. 7.1). While it is not practical to alter the time-step routinely, it is possible simply to remove the second corrector step, thus assessing the affect of essentially doubling the interval of solution. This change left the model results for the standard run virtually unchanged (Fig. 77). The simpler scheme proved somewhat less accurate during the summer period of fast turnover, but no substantial difference resulted even then. As mentioned earlier (Sect. 7.1), the exponential evaluation of the instantaneous rates determined the exact integral during whatever time interval the rates may be assumed to remain constant. This method is apparently appropriate for the ecological rates dealt with here.

12. Applications and Limitations

12.1 Ecosystem Modeling and Environmental Management

It has frequently been suggested that numerical ecosystem models should be applied directly to problems of environmental management as a tool for predicting the response of natural systems to perturbations and modifications of various kinds. Our feeling is that while such applications may be instructive, the results must be interpreted with caution and a healthy skepticism. There are, after all, many differences between model systems and real ecosystems. At best, the Narragansett Bay model is designed to represent our current understanding of the bay in its present state as described in Chapter 1. If the natural system is changed in some fundamental way, the conceptual model that lies behind all of the equations and computer programming may no longer apply. As it is presently designed, for example, the model cannot change to different system states that may result from large increases in organic inputs, drastic changes in salinity, anoxic bottom waters, etc. This is not to say that such restructuring cannot be modeled, perhaps even by modifications of the original program. But no model can reasonably be expected to deal with all possibilities, and thus may fall far short of representing or anticipating an event in the natural system.

Since many management decisions are tied to questions of large and/or long-term perturbations of the system, it may not be appropriate to extrapolate very far into the future using a model that is constrained by being closed and nonevolving. The real system is coupled by tidal fluxes, river flows, and other inputs to a vast and infinitely varied assortment of potential compartments. It is also made up of parts that adapt and evolve. New forms, new physiologies, and new interrelationships give it stability in its present state as well as the potential to change gradually over time, or more rapidly in response to strong perturbations.

These constraints become less important if the model is used to explore the responses of the present system to relatively small changes in parameters and processes that are specifically included in its formulation. While simulations of this kind are part of the sensitivity analysis described in the previous chapter, it is also possible to manipulate the model in ways that may yield information that can contribute to resolving management issues. For example, several years ago an electric utility company proposed to construct a nuclear-powered generating plant on the lower West Passage of Narragansett Bay, in the region of our model element 4 (Fig. 1). Part of the concern over the environmental impact of such a plant centered on the mortality of phytoplankton and zooplankton that would result

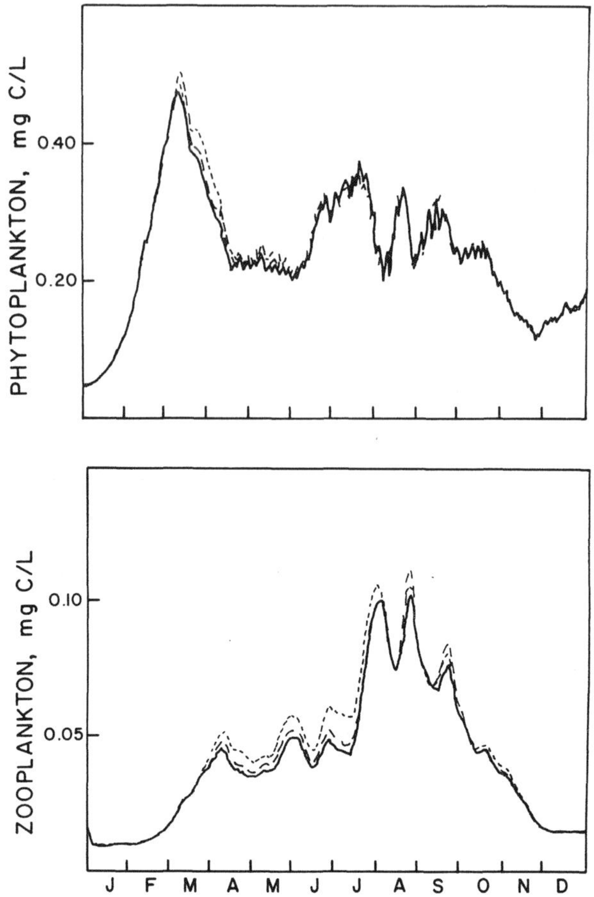

Fig. 78. Simulated phytoplankton and zooplankton population levels in element 4 of Narragansett Bay (Fig. 1) with a power plant cooling system in the same element causing 100% mortality of both populations in water passing through the plant. Simulations in which the flow rate of cooling water ranged from 1500 cfs (42.5 m^3/s; *dashed line*) to 4500 cfs (127.5 m^3/s; *dotted line*) are compared with the standard run in which there was no power plant *(solid line)*

from the use of large amounts of bay water for cooling. Without a simulation model it was difficult to anticipate what effect various mortality rates might have on the standing crop of plankton in the bay. The system is far too complicated to resolve intuitively, though preliminary answers to this particular question can be obtained rather easily with the model. Results of simulations using low and high estimates of the flow of bay water through the plant assuming 100% mortality suggest that there would be little, if any, change in the standing crop of phytoplankton or zooplankton, even in the general area around the power plant (Fig. 78). In fact, there is a surprising indication that standing crops may actually be higher at certain times of the year. The mechanism for such a response lies in the rapid conversion of nutrients from organic to inorganic form at times when other environmental conditions are more favorable for production than they were when the nutrients were originally fixed in organic form. We emphasize, however, that the model does not pretend to foretell the future; it has only done a series of complex calculations based on the relationships and parameter values that we have specified. It must be understood that the present model cannot tell us anything about spatial effects smaller than one element, nor can it be used to anticipate the effects of behavioral changes, species substitutions, or many other aspects of environmental change that may be associated with power plant construction and operation. But it can suggest possibilities which may not have been suspected otherwise, and it allows us to investigate the ramifications of any assumption we make.

Similarly, the Narragansett Bay model can be used to begin to explore other aspects of management activity. One area of increasing concern involves the tertiary treatment of urban sewage to remove nutrients. Since nutrient dynamics are described in some detail in this model, it may provide insight into the role of sewage inputs in influencing plankton dynamics in the system. If the model is run with nutrient input from sewage halved or doubled, the primary effect over most of the bay appears in the intensity of the winter–spring phytoplankton bloom (Fig. 79). For the upper bay and Providence River (elements 1 and 2), in contrast, there seems to be little effect of increased loading on the winter–spring bloom. Instead, increased nutrient input in this region more dramatically enhanced the summer phytoplankton bloom.

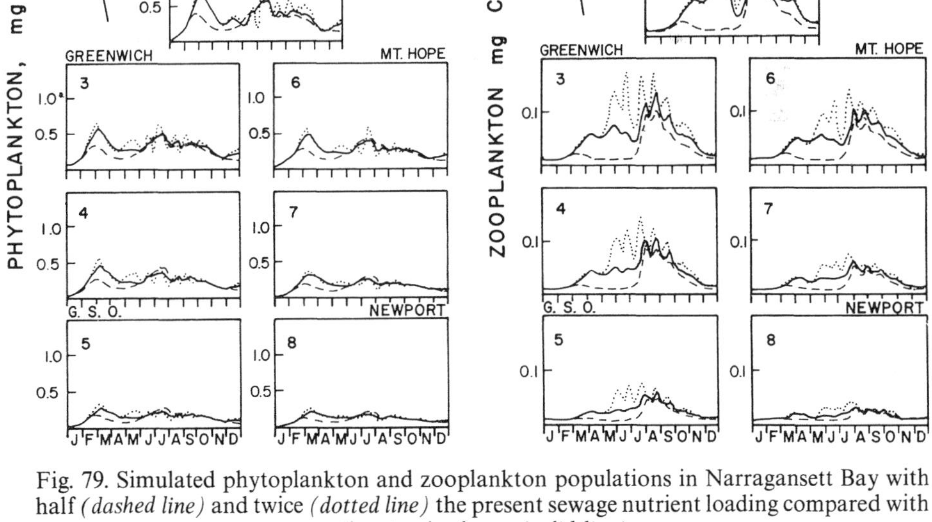

Fig. 79. Simulated phytoplankton and zooplankton populations in Narragansett Bay with half *(dashed line)* and twice *(dotted line)* the present sewage nutrient loading compared with the standard run *(solid line)*

The lowering of sewage nutrients and the associated phytoplankton changes markedly depressed the production of zooplankton until the late summer and fall (Fig. 79). The increased nutrient loading, however, was reflected in a larger standing crop of animals beginning in May and in marked phytoplankton–zooplankton cycles of relatively large amplitude (Fig. 79). Regardless of the changes in sewage nutrient inputs, the fall and early winter patterns of plankton dynamics appear to be shifted little from present conditions. The presence of large seasonal changes in forcing functions of temperature and light may be an important feature of temperate systems in that they serve as "narrow pass filters" to dampen or extinguish oscillations from year to year. This feature is often absent from many less mechanistic models which emphasize population interactions and then proceed to theoretical discussions of ecosystem stability (Rosenzweig, 1971; Kremer, in press).

Unfortunately, however, no matter how realistic the model, the results of simulations do not necessarily suggest the best management strategy. Even if we knew that the model simulations were absolutely correct about what would happen if the sewage input were removed or doubled, it is not clear if either alternative is more desirable than the present situation. The model offers no escape from value judgements and long links of supposition for those involved in management issues.

It is important to reemphasize that the interpretation of model results in management, and other applications must be closely restricted to effects and relationships deliberately included in the model. For this reason, it is most desirable that the management objectives and related questions to be posed to the model be specified from the outset. Only then can maximum information and confidence be achieved.

It would not be fair to end what has been, we hope, an honest but serious discussion of the development and application of numerical ecosystem models in general, and the Narragansett Bay model in particular, without mentioning that they can also be used for fun. Once the model is running, it is naturally tempting to play at least a few "what if" games. While somewhat related to the applications discussed earlier, such games are motivated more by fanciful curiosity than by the original modeling goals or possible management implications.

For example, what might Narragansett Bay be like if it had developed just as it is, except for its geographic location? What would the plankton dynamics be like if the bay were located in Nova Scotia, or further south near Chesapeake Bay, or still further south to Biscayne Bay, Florida? The results of substituting the appropriate light and temperature forcing functions for each of these areas in the standard run of the model were striking and a bit puzzling (Fig. 80). As the system is moved south from Nova Scotia, phytoplankton–zooplankton oscillations of increasing frequency and amplitude spread out from the summer period until they characterize the entire annual cycle. After analyzing the runs carefully, it became clear that this behavior resulted primarily from the effect of temperature on the time delay terms in the zooplankton population dynamics. As discussed in Chapter 5, both egg hatching and juvenile growth are temperature dependent. If temperatures remain high, the delay in converting assimilated phytoplankton into adult zooplankton biomass becomes very small, a condition characteristic of most models which do not include age structure in the formulation of the zooplankton compartment. The lesson from the simulation is not that the model does a poor job of simulating tropical

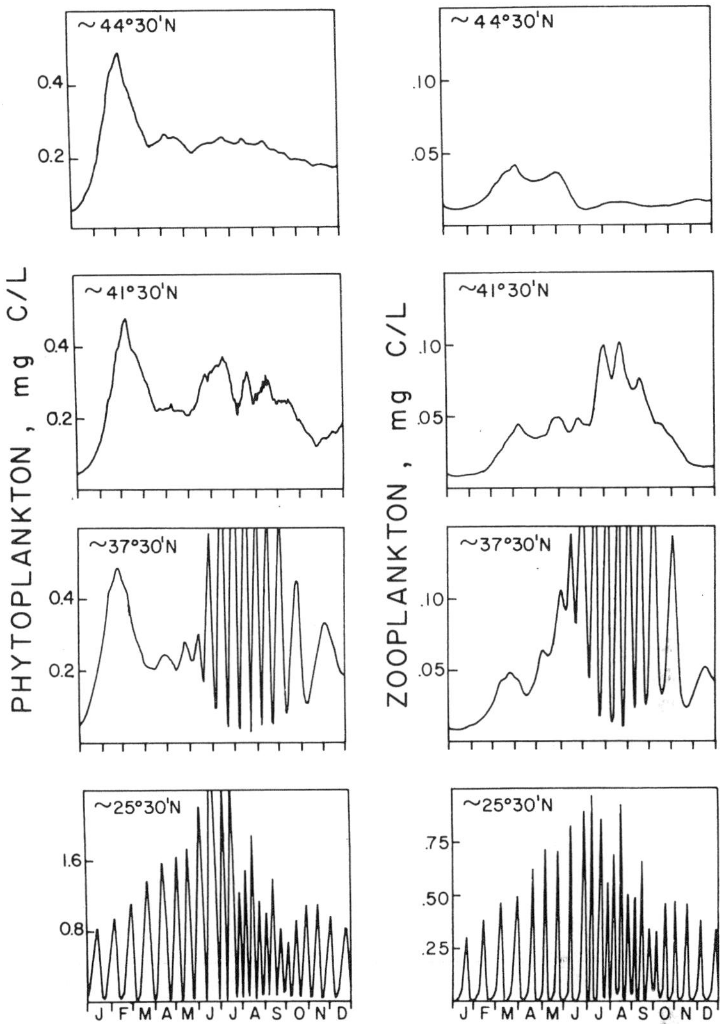

Fig. 80. Simulated phytoplankton and zooplankton populations in the Narragansett Bay model as it is moved down the Atlantic coast from approximately the latitude of the Bay of Fundy through Rhode Island and the Chesapeake Bay region to Biscayne Bay, Florida. In each case, all factors were the same as in the standard run (Rhode Island case) except for the annual cycles of solar radiation and temperature, which were adjusted to match empirical data for each area

zooplankton, but that the addition of the time delay terms associated with increased biological complexity contributes to damping the predator–prey interactions of both the model and, perhaps, Narragansett Bay. Perhaps in more oligotrophic tropical systems, where standing crops of plankton are much lower, time lags are less important to maintain stable populations, or the oscillations are practically unmeasurable.

12.2 Holism and Reductionism

We began our discussion of mechanistic numerical ecosytem models in Chapter 1 with the observation that projects like the Narragansett Bay model provide a mechanism for synthesizing a great many data on individual parts of the system. In this way, such models are tests of the reductionist philosophy that has dominated much of biological and ecological research. At present, there is disagreement about whether or not the emergent properties of a complex system such as Narragansett Bay can be simulated in a mechanistic fashion by combining data that have been collected on isolated parts of various marine systems in laboratories around the world. While we agree with those who point out that the holistic principle applies to living systems (von Bertalanffy, 1968) and that natural communities are more than, or at least something other than, the sum of their parts (Odum, 1971; Patten, 1971; Mann, 1972, 1975), we also feel that modeling work has made, and will continue to make, important contributions to our understanding of ecological systems. Like "all the king's horses and all the king's men", we may never be able to fit everything back together again, but there is much to be learned from the attempt.

In addition to all of the limitations we have already discussed, it is worth emphasizing again that the Narragansett Bay model, like almost all other mechanistic models, is largely deterministic. With the exception of the stochastic cloud factor (Chap. 3), the model will always give the same simulation for the same initial conditions, forcing functions and parameter values. It is of interest that physics, a discipline sometimes enviously regarded by biologists as ultimately deterministic, effectively reached an impass which was only successfully broached by a rejection of the deterministic approach. It was the concession that absolute predictability of any physical process or attribute was impossible, regardless of the completeness of our information, which provided the breakthrough. It became axiomatic that natural variability exists at every level of resolution and statistical probabilities for behavior became the paradigm for continued progress. Ecologists are still largely occupied with the pursuit of certainty in their measurements and in most of their models. Future efforts in both areas will no doubt include considerations of stochastic processes as well as a continuing inquiry into the role of uncertainty in natural systems.

Acknowledgements. The development of a mechanistic model requires a synthesis of effort and ideas as well as data. We are grateful to the many friends, colleagues and students who shared all three with us. Almost all of the people in our own ecological systems group played some part in developing the model. Sharon Northby, Fred Short, Eric Klos, Pat Kremer, and Ann Durbin were responsible for collection and analysis of many of the verification data. Pat Kremer, Ann Durbin and Candace Oviatt provided unpublished data and helped with a number of formulations. Fred Short, Wendell Hahm and Terry Smith assisted with computer programming and some of the plotting, and many people at the URI computer center frequently went out of their way to accommodate us. Members of the Ocean Engineering Department at URI, including Frank White, Kurt Hess, Malcolm Spaulding and George Brown, made it possible for us to include hydrodynamic effects in the model, and worked with us on a variety of problems from numerical analysis to data collection. Ted Smayda, Ted Durbin, Gabe Vargo, Gary Hitchcock, and Jim Yoder in the phytoplankton ecology group in Oceanography often provided us with data, suggestions, and valuable criticisms. Jan Northby in Physics criticized the discussion of mathematical concepts. Harlan Lampe in Resource Economics and Nelson Marshall and Perry Jeffries in Oceanography were instrumental in the early phases of the project. Figures for the book were prepared by Mickey Leonard and Betsy Watkins and the text was typed by Germaine Webb. The research was supported by the Office of Sea Grant Programs, NOAA, U.S. Dept. of Commerce.

We also received help in a variety of forms from people outside the University of Rhode Island, including Don Heinle at the University of Maryland, Ed Zillioux at the University of Miami, H. T. Odum at the University of Florida, Geoffrey Laurence at the National Marine Fisheries Service, Narragansett, and the Marine Fisheries group of the R.I. Department of Natural Resources, Wickford.

While the development of the model was a stimulating and rewarding effort, it was not without periods of difficulty and discouragement, and we are especially grateful to our wives, Patricia Kremer and Pendleton Nixon, to John Knauss, Dean of the Graduate School of Oceanography, to Niels Rorholm, Coordinator of the Sea Grant Program at URI, and to our colleague and friend, Candace Oviatt, for constant encouragement and support throughout the project.

References

Anraku, M.: Influences of the Cape Cod Canal on the hydrography and on the copepods in Buzzards Bay and Cape Cod Bay, Massachusetts. II. Respiration and feeding. Limnol. Oceanogr. **9**, 195–206 (1964)

Bannister, T.T.: Production equations in terms of chlorophyll concentration, quantum yield, and upper limit to production. Limnol. Oceanogr. **19**, 1–12 (1974)

Bannister, T.T.: A general theory of steady state phytoplankton growth in a nutrient saturated mixed layer. Limnol. Oceanogr. **19**, 13–30 (1974a)

Bannister, T.T.: Reply to comment by G. A. Riley. Limnol. Oceanogr. **20**, 152–153 (1975)

Beklemishev, C.W.: Superfluous feeding of marine herbivorous zooplankton. Rapp. P.-V. Réun., Cons. Perm. Int. Explor. Mer. **153**, 108–113 (1962)

Berman, M.S., Richman, S.: The feeding behavior of *Daphnia pulex* from Lake Winnebago, Wisc. Limnol. Oceanogr. **19**, 105–109 (1974)

Bertalanffy, L. von: Quantitative laws in metabolism and growth. Quart. Rev. Biol. **32**, 217–231 (1957)

Bertalanffy, L. von: General System Theory. New York: George Braziller Inc. 1968, 289 pp.

Bigelow, H.B., Schroeder, W.C.: Fishes of the Gulf of Maine. Fishery Bull. of the Fish and Wildlife Service **74** (1953)

Blaxter, J.H.S.: The feeding of herring larvae and their ecology in relation to feeding. In: Calif. Cooperative Oceanic Fisheries Investigations Reports, 1965, Vol. X

Bowditch, N.: American Practical Navigator. Washington, D.C. U.S. Govt. Printing Office, U.S. Navy Hydrographic Office Pub. **9** (1958)

Burgis, M.J.: The effect of temperature on the development time of eggs of *Thermocyclops* sp., a tropical cyclopoid copepod from Lake George, Uganda. Limnol. Oceanogr. **15**, 742–747 (1970)

Burkholder, P.R., Doheny, T.E.: Proximate analysis and amino acid composition of some marine phytoplankton and bacteria. Mar. Lab., Town of Hempstead, N.Y., Dept. of Conservation and Waterways. **8** (1972)

Butler, E.I., Corner, E.D.S., Marshall, S.M.: On the nutrition and metabolism of zooplankton. VI. Feeding efficiency of *Calanus* in terms of nitrogen and phosphorus. J. Mar. Biol. Assoc. U. K. **49**, 977–1001 (1969)

Butler, E.I., Corner, E.D.S., Marshall, S.M.: On the nutrition and metabolism of zooplankton. VII. Seasonal survey of nitrogen and phosphorus excretion by *Calanus* in the Clyde Sea area. J. Mar. Biol. Assoc. U. K. **50**, 525–560 (1970)

Callen, H.B.: Thermodynamics. New York: Wiley 1963, Chaps. 1, 2

Canale, R.P., Hineman, D.F., Nachiappan, S.: A biological production model for Grand Transverse Bay. Univ. Mich., Sea Grant Prog. Tech. Rep. **37** (1974)

Carpenter, E.J., Guillard, R.R.L.: Intraspecific differences in nitrate half-saturation constants for three species of marine phytoplankton. Ecology **52**, 183–185 (1971)

Collins, B.P.: Suspended material transport in lower Narragansett Bay and Western R. I. Sound. M. S. Thesis, Univ. R. I., Kingston, Rhode Island (1974)

Comita, G.W., Comita, J.J.: Egg production in *Tigriopus brevicornis*. Some Contemp. Stud. Mar. Sci. 171–185 (1966)

Conover, R.J.: Oceanography of Long Island Sound. VI. Biology of *Acartia clausi* and *A. tonsa*. Bull. Bing. Ocean. Coll. **15**, 156–233 (1956)

Conover, R.J.: Regional and seasonal variation in the respiration rate of marine copepods. Limnol. Oceanogr. **4**, 259–268 (1959)

Conover, R.J.: The feeding behavior and respiration of some marine planktonic crustacea. Biol. Bull. **119**, 399–415 (1960)

Conover, R.J.: Food relations and nutrition of zooplankton. Exp. Mar. Ecol., Narr. Mar. Lab., Univ. R. I., Kingston, Rhode Island. Occasional Publ. No. 2. Feb. (1964)

Conover, R.J.: Assimilation of organic matter by zooplankton. Limnol. Oceanogr. **11**, 338–345 (1966)

Conover, R.J.: Factors affecting the assimilation of organic matter by zooplankton and the question of superfluous feeding. Limnol. Oceanogr. **11**, 346–354 (1966a)

Conover, R.J.: Zooplankton—Life in a nutritionally dilute environment. Am. Zoologist **8**, 107–118 (1968)

Conover, R.J., Corner, E.D.S.: Respiration and nitrogen excretion by some marine zooplankton in relation to their life cycle. J. Mar. Biol. Assoc. U.K. **48**, 49–75 (1968)

Corner, E.D.S.: Laboratory studies related to zooplankton production in the sea. In: Edwards, R.W., Garrod, J.D. (eds.). Conservation and Productivity of Natural Waters. Symp. Zool. Soc. Lond. 29. New York: Academic 1972, pp. 185–201

Corner, E.D.S., Cowey, C.B., Marshall, S.M.: On the nutrition and metabolism of zooplankton. III. Nitrogen excretion by *Calanus*. J. Mar. Biol. Assoc. U.K. **45**, 429–442 (1965)

Corner, E.D.S., Davis, A.G.: Plankton as a factor in the nitrogen and phosphorus cycles in the sea. Advan. Mar. Biol. **9**, 101–204 (1971)

Corner, E.D.S., Newell, B.S.: On the nutrition and metabolism of zooplankton. IV. The forms of nitrogen excreted by *Calanus*. J. Mar. Biol. Assoc. U.K. **47**, 113–120 (1976)

Coughlan, J., Ansell, A.D.: A direct method for determining the pumping rate of siphonate bivalves. J. Con. Int. Explor. Mer. **29**, 205–213 (1964)

Cronin, J. (ed.): Estuarine Research. Proc. 2nd Int. Est. Res. Conf., Myrtle Beach, S.C. 1973. New York: Academic 1975, Vols. 1 and 2

Curl, H.: Analysis of carbon in marine plankton organisms. J. Mar. Res. **20**, 181–188 (1962)

Cushing, D.H.: On the nature of production in the sea. Fish. Invest., Ser. II, **22**, 1–40 (1959)

Cushing, D.H.: The work of grazing in the sea. In: Crisp, D.J. (ed.) Grazing in Terrestrial and Marine Environments. London: Blackwell Sci. Publ. 1964, pp. 207–225

Cushing, D.H.: Grazing by herbivorous copepods in the sea. J. Cons. Perm. Int. Explor. Mer **32**, 70–82 (1968)

Cushing, D.H.: A comparison of production in temperate seas and the upwelling areas. Proc. SANCOR Symp. Oceanography in South Africa. Durban, S. Africa (1970), pp. 1–26

Cushing, D.H., Nicholson, H.N.: Studies on a *Calanus* patch. IV. Nutrient salts of the north-east coast. J. Mar. Biol. Assoc. U.K. **43**, 373–386 (1963)

DiToro, D.M., O'Connor, D.J., Thomann, R.V.: A dynamic model of phytoplankton populations in the Sacramento-San Joaquin Delta. Advan. in Chem. Series **106**, 131–180 (1971)

Droop, M.R.: Nutrient limitation in osmotrophic protista. Am. Zoologist **13**, 209–214 (1973)

Dugdale, R.C.: Nutrient limitations in the sea: dynamics, identification, and significance. Limnol. Oceanogr. **12**, 685–695 (1967)

Dugdale, R.C., MacIsaac, J.J.: A computation model for the uptake of nitrate in the Peru upwelling region. Invest. Pesq. **35**, 299–308 (1971)

Durbin, A.G.: The role of fish migration in two coastal ecosystems: the Atlantic Menhaden *(Brevoortia tyrannus)* in Narragansett Bay, and the Alewife *(Alosa pseudoharengus)* in Rhode Island ponds. Ph. D. Thesis, Univ. R. I., Kingston, Rhode Island (1976)

Durbin, A.G., Durbin, E.G.: Grazing rates of Atlantic Menhaden *Brevoortia tyrannus*, as a function of particle size and concentration. Mar. Biol. **33**, 265–277 (1975)

Durbin, E.G.: Studies on the autecology of the marine diatom *Thalassiosira nordenskioldii* Cleve. I. The influence of daylength, light intensity, and temperature on growth. J. Phycol. **10**, 220–225 (1974)

Durbin, E.G., Krawiec, R.W., Smayda, T.J.: Seasonal studies on the relative importance of different size fractions of phytoplankton in Narragansett Bay (USA). Mar. Biol. **32**, 271–287 (1975)

Edmondson, W.T.: Food supply and reproduction of zooplankton in relation to phytoplankton populations. Rapp. P.-v. Réun., Cons. Perm. Int. Explor. Mer. **153**, 137–141 (1961)

Edmondson, W.T., Comita, G.W., Anderson, G.C.: Reproductive rate of copepods in nature and its relation to phytoplankton population. Ecology **43**, 625–634 (1962)

Eppley, R.W.: Temperature and phytoplankton growth in the sea. Fish. Bull. **70**, 1063–1085 (1972)

Eppley, R.W., Carlucci, A.F., Holm-Hansen, O., Kiefer, D., McCarthy, J.J., Venrich, E., Williams, P.M.: Phytoplankton growth and composition in shipboard cultures supplied with nitrate, ammonium, or urea as a nitrogen source. Limnol. Oceanogr. **16**, 741–751 (1971)

Eppley, R.W., Coatsworth, J.L.: Uptake of nitrate and nitrite by *Ditylum brightwelli*—kinetics and mechanisms. J. Phycol. **4**, 151–156 (1968)

Eppley, R.W., Rogers, J.N., McCarthy, J.J.: Half saturation constants for uptake of nitrate and ammonium by marine phytoplankton. Limnol. Oceanogr. **14**, 912–920 (1969)

Eppley, R.W., Rogers, J.N., McCarthy, J.J.: Light-dark periodicity in nitrogen assimilation of the marine phytoplanktors *S. costatum* and *C. huxleyi* in N-limited chemostat cultures. J. Phycol. **7**, 150–154 (1971)

Eppley, R.W., Strickland, J.D.H.: Kinetics of marine phytoplankton growth. Advan. Microbiol. of the Sea **1**, 23–62 (1968)

Eppley, R.W., Thomas, W.H.: Comparison of half-saturation constants for growth and nitrate uptake of marine phytoplankton. J. Phycol. **5**, 375–379 (1969)

Farrington, J.W., Quinn, J.G.: Petroleum hydrocarbons and fatty acids in waste water effluents. J. Water Pollution Cont. Fed. **45**, 704–712 (1973)

Fee, E.J.: A numerical model for the estimation of phytosynthetic production, integrated over time and depth, in natural waters. Limnol. Oceanogr. **14**, 906–911 (1969)

Fee, E.J.: A numerical model for determining integral primary production and its application to Lake Michigan. J. Fish. Res. Bd. Can. **30**, 1447–1468 (1973)

Fee, E.J.: Modelling primary production in water bodies: A numerical approach that allows vertical inhomogeneities. J. Fish. Res. Bd. Can. **30**, 1469–1473 (1973a)

Fish, C.J. (ed.). Physical oceanography of Narr. Bay-R.I. Sound. I.S.P. Final Report, Narr. Mar. Lab., Univ. R.I., Kingston, Rhode Island (1953)

Fleming, R.H.: The control of diatom populations by grazing. J. Cons. Perm. Expl. Mer. **14**, 210–227 (1939)

Fogg, G.E.: Algal Cultures and Phytoplankton Ecology. Madison, Wis.: Univ. Wisconsin Press 1965, 126 pp.

Frolander, H.F.: The biology of the zooplankton of the Narragansett Bay area. Ph. D. Thesis, Brown University, Providence, Rhode Island (1955)

Frost, B.W.: Effects of size and concentration of food particles on the feeding behavior of the marine planktonic copepod *Calanus pacificus*. Limnol. Oceanogr. **17**, 805–815 (1972)

Frost, B.W.: A threshold feeding behavior in *Calanus pacificus*. Limnol. Oceanogr. **20**, 263–266 (1975)

Fuhs, G., Demmerle, S., Canelli, E., Chen, M.: Characterization of P-limited plankton algae. Nutrients and Eutrophication. ASLO Special Symp. 1972, Vol. I, 113–132

Gauld, D.T.: Grazing rate of planktonic copepods. J. Mar. Biol. Assoc., U.K. **29**, 695–706 (1951)

Gauld, D.T., Raymont, J.E.G.: The respiration of some planktonic copepods. II. The effect of temperature. J. Mar. Bio. Assoc., U.K. **31**, 447–460 (1953)

Gerber, R.P., Marshall, N.: Ingestion of detritus by the lagoon pelagic community at Eniwetok Atoll. Limnol. Oceanogr. **19**, 815–824 (1974)

Ghittino, P.: The diet and general fish husbandry. In: Halver, J.E. (ed.). Fish Nutrition. New York: Academic 1972

Goldberg, E.D., McCave, I.N., O'Brien, J.J., Steele, J.H. (eds.): Marine Modeling. The Sea. New York: Wiley-Interscience 1977, Vol. VI, 1048 pp.

Goldman, J.C., Carpenter, E.J.: A kinetic approach to the effect of temperature on algal growth. Limnol. Oceanogr. **19**, 756–766 (1974)

Grenney, W.J., Bella, D.A., Curl, H.C., Jr.: A theoretical approach to interspecific competition in phytoplankton communities. Am. Naturalist. **107**, 405–425 (1973)

Greze, V.N., Baldina, E.P.: Population dynamics and annual production of *Acartia clausi* Giesbr. and *Centropages kroyeri* Giesbr. in the neritic zone of the Black Sea. (Fish. Res. Bd. Can. Trans. 893, 1967) Trudy Sevastopol'skoi Biologicheskoi stantsii, Akad. Nauk Ukrain. SSR **17**, 249–261 (1964)

Hale, S.S.: The role of benthic communities in the nutrient cycles of Narragansett Bay. M.S. Thesis, Univ. R.I., Kingston, Rhode Island (1974)

Hall, C., Day, J. (eds): Ecosystem Modeling in Theory and Practice: An Introduction with Case Histories. New York: John Wiley and Sons 1977

Hamwi, A., Haskins, H.H.: Oxygen consumption and pumping rates in the hard clam *Mercenaria mercenaria*: a direct method. Science **163**, 823–824 (1969)

Hanton, J.T.: Algal phosphate uptake kinetics: growth rates and limiting phosphate concentrations. M.S. Thesis, Univ. North Carolina, Chapel Hill, (1969)

Haq, S.M.: Nutritional physiology of *Metridia lucens* and *M. longa* from the Gulf of Maine. Limnol. Oceanogr. **12**, 40–51 (1967)

Hargrave, B.T., Geen, G.H.: Phosphorus excretion by zooplankton. Limnol. Oceanogr. **13**, 332–343 (1968)

Harris, E.: The nitrogen cycle in Long Island Sound. Bull. Bing. Oceanogr. Coll. **17**, 31–65 (1959)

Harris, J.G.K.: A mathematical model describing the possible behavior of a copepod feeding continuously in a relatively dense randomly distributed population of algal cells. J. Cons. Perm. Int. Explor. Mer. **32**, 83–92 (1968)

Heinle, D.R.: Production of a calanoid copepod, *Acartia tonsa*, in the Pawtuxent River Estuary. Ches. Sci. **7**, 59–74 (1966)

Heinle, D.R.: The effects of temperature on the population dynamics of estuarine copepods. Ph. D. Thesis, Univ. Maryland, College Park (1969)

Heinle, D.R.: Temperature and zooplankton. Ches. Sci. **10**, 186–209 (1969a)

Heinle, D.R., Flemer, D.A., Ustach, J.F., Murtagh, R.A., Harris, R.P.: The role of organic debris and associated micro-organisms in pelagic estuarine food chains. Tech. Rept. 22. Univ. Maryland (Natl. Res. Inst. Ref. No. 72–29), 1973

Herman, S.S.: The planktonic fish eggs and larvae of Narragansett Bay. M. S. Thesis, Univ. R. I., Kingston, Rhode Island (1958)

Hess, K.W.: A 3-dimensional numerical model of steady gravitational circulation and salinity distribution in Narragansett Bay. Ph. D. Thesis, Univ. R. I., Kingston, Rhode Island (1974)

Hess, K., White, F.: A numerical tidal model of Narragansett Bay. Sea Grant Mar. Tech. Rept. **20**, Univ. R.I., Kingston, Rhode Island (1974)

Hicks, S.D.: The physical oceanography of Narragansett Bay. Limnol. Oceanogr. **4**, 316–327 (1959)

Hicks, S.D., Frazier, D.E., Garrison, L.E.: Physical oceanographic effects of proposed hurricane protection structures on Narragansett Bay under normal conditions. Ref. No. 56, 12. Hurricane Protection Project, 1956

Hitchcock, G., Smayda, T.J.: The importance of light in the initiation of the 1972–73 winter-spring diatom bloom in Narragansett Bay. Limnol. Oceanogr., **22**, 126–131 (1977)

Hodgkin, E.P., Rippingale, R.J.: Interspecies conflict in estuarine copepods. Limnol. Oceanogr. **16**, 573–576 (1971)

Holling, C.S.: The functional response of invertebrate predators to prey density. Mem. Ent. Soc. Can. **48**, 1–85 (1966)

Holling, C.S., Ewing, S.: Blind-mans bluff: exploring the response surface generated by a realistic ecological simulation model. Proc. Int. Sym. Stat. Ecol., New Haven, Connecticut, 1969, 70 pp.

Hulsizer, E.E.: Zooplankton of Lower Narragansett Bay 1972–73. Ches. Sci. **17**, 260–270 (1976)

Ignatiades, L., Smayda, T.J.: Autecological studies on the marine diatom *Rhizosolenia fragilissima* Bergon. I. The influence of light, temperature, and salinity. J. Phycol. **6**, 332–339 (1970)

Ikeda, T.: Relationship between respiration rate and body size in marine plankton animals as a function of the temperature of habitat. Bull. Fac. Fish. Hokkaido Univ. **21**, 91–112 (1970)

Ivlev,V.S.: The biological productivity of waters. Uspekhi Sovrem. Biol. **19**, 98–120 (1945)
Jaworski,N.A., Lear,D.W., Villa,O., Jr.: Nutrient management in the Potomac Estuary. Nutrients and Eutrophication. ASLO Special Symp. 1972, Vol. I, pp. 246–272
Jeffries,H.P.: Succession of two *Acartia* species in estuaries. Limnol. Oceanogr. **7**, 354–364 (1962)
Jeffries,H.P.: Diets of juvenile Atlantic menhaden *(Brevoortia tyrannus)* in three estuarine habitats as determined from fatty acid composition of gut contents. J. Fish. Res. Bd. Can. **32**, 587–592 (1975)
Jeffries,H.P., Johnson,W.C.: Distribution and abundance of zooplankton. In: Coastal and Offshore Environmental Inventory Cape Hatteras to Nantucket Shoals. Marine Pub. Ser. No. 2, Univ. R.I., Kingston, Rhode Island 1973
Jitts,H.R., McAllister,C.D., Stephens,K., Strickland,J.D.H.: The cell division rates of some marine phytoplankters as a function of light and temperature. J. Fish. Res. Bd. Can. **21**, 139–157 (1964)
Ketchum,B.H.: The absorption of phosphate and nitrate by illuminated cultures of *Nitzschia closterium*. Am. J. Botany. **26**, 399–406 (1939)
Ketchum,B.H.: Regeneration of nutrients by zooplankton. Cons. Int. Explor. Mer **153**, 142–146 (1962)
Kremer,J.N.: An analysis of the stability characteristics of an estuarine ecosystem model. Proc. Marsh-Estuarine Simulation Symp., Bell Baruch Inst. Mar. Res., Univ. S. Carolina, in press
Kremer,J.N., Nixon,S.W.: An ecological simulation model of Narragansett Bay — the plankton community. In: Cronin,J. (ed.). Estuarine Research. Proc. 2nd Int. Est. Res. Conf., Myrtle Beach,S.C. New York: Academic 1975, Vol. 1, pp. 672–690
Kremer,P.: The ecology of the ctenophore, *Mnemiopsis leidyi* in Narragansett Bay. Ph. D. Thesis, Univ. R. I., Kingston, Rhode Island (1975)
Kremer,P.: Nitrogen regeneration by the ctenophore, *Mnemiopsis leidyi*. In: Howell,F.G., Gentry,J.B., Smith,M.H. (eds.) Mineral cycling in southeastern ecosystems. ERDA Symp. Ser. (Conf-740513), 1975a, pp. 279–290
Kremer,P.: Population dynamics and ecological energetics of a pulsed zooplankton predator, the ctenophore, *Mnemiopsis leidyi*. Estuarine Processes. Proc. 3rd Int. Estuarine Res. Conf., Galveston, Texas. New York: Academic 1976, Vol. 1, pp. 197–215
Kremer,P., Nixon,S. Distribution and abundance of the ctenophore, *Mnemiopsis leidyi* in Narragansett Bay. Est. Coast. Mar. Sci., **4**, 627–639 (1976)
Lampe,H.L., Nixon,S.W.: A bioeconomic model of Narragansett Bay. Paper presented at the 33rd Annual Meeting, ASLO, Kingston, R. I. (1970)
Landry,M.R.: Seasonal temperature effects and predicting development rates of marine copepod eggs. Limnol. Oceanogr. **20**, 305–496 (1975)
Lassiter,R.R., Hayne,D.W.: A finite difference model for simulation of dynamic processes in ecosystems. In: Patten,B.C. (ed.). Systems Analysis and Simulation in Ecology. New York: Academic 1971, Vol. I, pp. 367–440
Laurence,G.C.: Experimental growth and metabolism of winter flounder *Pseudopleuronectes americanus*, from hatching through metamorphosis at three temperatures. Mar. Biol. **33**, 223–230 (1975)
Leendertse,J.J.: Aspects of a computational model for longperiod wave propagation. Mem. RM 5294-PR, Rand Corp., Santa Monica, Calif., 1967
Lehman,J.T., Botkin,D.B., Likens,G.E.: The assumptions and rationales of a computer model of phytoplankton population dynamics. Limnol. Oceanogr. **20**, 343–364 (1975)
Levin,S.A. (ed.): Ecosystem Analysis and Prediction. Proc. SIAM-SIMS Conf., Alta, Utah 1975
Levine,E.: The tidal energetics of Narragansett Bay. M. S. Thesis, Univ. R. I., Kingston, Rhode Island (1972)
Lewin,J.C.: The dissolution of silica from diatom walls. Geochimica et Cosmochimica Acta **21**, 182–198 (1961)
Loosanoff,V.L.: Effect of temperature upon shell movements of clams, *Venus mercenaria* (L). Biol. Bull. **76**, 171–182 (1939)
Lorenzen,C.J.: Extinction of light in the ocean by phytoplankton. J. Cons. Int. Explor. Mer **34**, 262–267 (1972)

Lotka, A. J.: Elements of physical biology. Baltimore: Williams and Wilkins 1925. (Reprinted as: Elements of Mathematical Biology. New York: Dover 1956)
Loveland, R., Chu, D.: Oxygen consumption and water movement in *Mercenaria mercenaria*. Compt. Biochem. Physiol. **29**, 173–184 (1969)
Lowry, W. P.: Weather and Life. New York: Academic, 1967, 305 pp.
MacIsaac, J. J., Dugdale, R. C.: The kinetics of nitrate and amonia uptake by natural populations of marine phytoplankton. Deep-Sea Res. **16**, 45–57 (1969)
MacIsaac, J. J., Dugdale, R. C.: Interactions of light and inorganic nitrogen in controlling nitrogen uptake in the sea. Deep-Sea Res. **19**, 209–232 (1972)
McAllister, C. D.: Zooplankton rations, phytoplankton mortality, and the estimation of marine production. In: J. H. Steele (ed.) Marine Food Chains. Berkeley: Univ. Cal. 1970, pp. 419–457
McAllister, C. D., Shah, N., Strickland, J. D. H.: Marine phytoplankton photosynthesis as a function of light intensity: a comparison of methods. J. Fish. Res. Bd. Can. **21**, 159–181 (1964)
McCarthy, J. J.: The uptake of urea by marine phytoplankton. J. Phycol. **8**, 216–222 (1972)
McLaren, I. A.: Some relationships between temperature and egg size, body size, development rate, and fecundity of the copepod *Pseudocalanus*. Limnol. Oceanogr. **10**, 528–538 (1965)
McLaren, I. A.: Predicting development rate of copepod eggs. Biol. Bull. **131**, 457–469 (1966)
McLaren, I. A., Corkett, C. J., Zillioux, E. J.: Temperature adaptation of copepod eggs from the Arctic to the Tropics. Unpubl. ms., Nat. Mar. Water Qual. Lab, EPA, Narragansett, Rhode Island (1969)
McMaster, R. L.: Sediments of Narragansett Bay systems and Rhode Island Sound, Rhode Island. J. Sed. Petrol. **30**, 249–274 (1960)
Mann, K. H.: The analysis of aquatic ecosystems, pp. 1–14. In: Clark, R. B., Wooton, R. J. (eds.) Essays in Hydrobiology. Exeter, U. K.: Univ. Exeter 1972
Mann, K. H.: Relationship between morphometrie and biological functioning in three coastal inlets of Nova Scota. In: Cronin, E. (ed.). Estuarine Research. Proc. 2nd Int. Est. Res. Conf., Myrtle Beach, S. C. New York: Academic 1975, Vol. I, pp. 634–644
Marchessault, G. D.: The application of delayed recruitment models to two commercial fisheries. M. S. Thesis, Univ. R. I., Kingston, Rhode Island (1974)
Margalef, R. A.: Some critical remarks on the usual approaches to ecological modeling. Inv. Pesq. **37**, 621–640 (1973)
Marshall, S. M., Nicholls, A. G., Orr, A. P.: On the biology of *Calanus finmarchicus*. VI. Oxygen consumption in relation to environmental conditions. J. Mar. Biol. Assoc. U. K. **20**, 1–27 (1935)
Marshall, S. M., Orr, A. P.: On the biology of *Calanus finmarchicus*. VII. Factors affecting egg production. J. Mar. Bio. Assoc. U. K. **30**, 527–548 (1952)
Marshall, S. M., Orr, A. P.: Experimental feeding of the copepod *Calanus finmarchicus* on phytoplankton cultures labelled with radioactive carbon. Pap. Mar. Biol. Ocean., Deep-Sea Res., Suppl. to Vol. **3**, 110–114 (1955)
Marshall, S. M., Orr, A. P.: On the biology of *Calanus finmarchicus*. VIII. Food uptake, assimilation and excretion in adult and stage V. *Calanus*. J. Mar. Bio. Assoc. U. K. **34**, 495–529 (1955a)
Martin, J. H.: Phytoplankton-zooplankton relationships in Narragansett Bay. Limnol. Oceanogr. **10**, 185–191 (1965)
Martin, J. H.: Phytoplankton-zooplankton relationships in Narragansett Bay. II. The seasonal importance of zooplankton grazing excretion. Ph. D. Thesis, Univ. R. I., Kingston, Rhode Island (1966)
Martin, J. H.: Phytoplankton-zooplankton relationships in Narragansett Bay. III. Seasonal changes in zooplankton excretion rates in relation to phytoplankton abundance. Limnol. Oceanogr. **13**, 63–71 (1968)
Matthiessen, G. C.: Rome Point investigations, Narragansett Bay ichthyo-plankton survey. Final Report of Mar. Res. Inc., Falmouth, Mass. to Narragansett Electric Co., 68 pp. (1974)
Mayzaud, P.: Respiration and excretion of zooplankton. II. Studies of the metabolic characteristics of starved animals. Mar. Bio. **21**, 19–28 (1973)

Michaelis, L., Menten, M. L.: Der Kinetic der Inverteinwirkung. Biochem. Z. **49**, 333–369 (1913)

Mitchell-Innes, B. A.: Ecology of the phytoplankton of Narragansett Bay and the uptake of silica by natural populations and the diatoms *Skeletonema costatum* and *Detonula confervacea*. Ph. D. Thesis, Univ. R. I., Kingston, Rhode Island (1973)

Monod, J.: Recherches sur la croissance des cultures bacteriennes. Paris: Herman et Cie, (1942)

Morris, I., Beardall, J.: Photorespiration in marine phytoplankton, In: Cech, J. J., Jr., Bridges, D. W., Horton, D. B. (eds.). Respiration of Marine Organisms. South Portland, Maine: TRIGOM, 1975, pp. 11–23

Morton, R. W.: Spatial and temporal distribution of suspended sediment; Narragansett Bay and Rhode Island Sound. In: Nelson, B. W. (ed.). Environmental framework of coastal plain estuaries. Geol. Soc. Am. Mem. 133, 1972

Mullin, M. M.: Some factors affecting the feeding of marine copepods of the genus *Calanus*. Limnol. Oceanogr. **8**, 239–250 (1963)

Mullin, M. M.: Production of zooplankton in the ocean: the present status and problems. Oceanogr. Mar. Bio. Ann. Rev. **7**, 293–314 (1969)

Mullin, M. M., Brooks, E. R.: Growth and metabolism of two planktonic marine copepods as influenced by temperature and type of food. In: Steele, J. H. (ed.). Marine Food Chains. Berkeley: Univ. Calif. 1970, pp. 74–95

Mullin, M. M., Stewart, E. F., Fuglister, F. J.: Ingestion of planktonic grazers as a function of concentration of food. Limnol. Oceanogr. **20**, 259–262 (1975)

Nihoul, J. C. J. (ed.). Modeling of marine systems. Elsevier Oceanography Series No. 10. New York: Elsevier Sci. Publ. Co., 1975, 272 pp.

Nixon, S. W., Kremer, J. N.: Narrangansett Bay—the development of a composite simulation model for a New England esturary. In: Ecosystem Modeling in Theory and Practice: An Introduction with Case Histories. Hall, C., Day, J. (eds.). New York: Wiley 1977, pp. 622–673.

Nixon, S. W., Kremer, J. N., Oviatt, C.: Narragansett Bay systems ecology program, Data Report, in preparation

Nixon, S. W., Oviatt, C. A.: Analysis of local variation in the standing crop of *Spartina alterniflora*. Bot. Marina. **16**, 103–109 (1973)

Nixon, S. W., Oviatt, C. A.: Ecology of a New England salt marsh. Ecol. Monographs **43**, 463–498 (1973a)

Nixon, S. W., Oviatt, C. A., Hale, S. S.: Nitrogen regeneration and the metabolism of coastal marine bottom communities. In: Anderson, J., Macfayden, A. (eds.). The Role of Terrestrial and Aquatic Organisms in Decomposition Processes. London: Blackwell Sci. Pub., 1976, pp. 269–283

Nixon, S. W., Oviatt, C. A., Kremer, J. N., Perez, K.: The use of numerical models and laboratory microcosms in estuarine ecosystem analysis—simulations of a winter phytoplankton bloom. Proc. Marsh–Estuarine Simulation Symp., Bell Baruch. Columbia: Univ. S. Carolina Press 1977

North, W. J., Integration of environmental conditions by a marine organism. In: Olson, T. A., Burgess, F. J. (eds.). Pollution and Marine Ecology. New York: Wiley 1967, pp. 195–224

O'Brien, J. J., Wroblewski, J. S.: An ecological model of the lower marine trophic levels on the continental shelf off West Florida. Technical Report, Geophys. Fluid Dyn. Inst., Florida St. Univ., Tallahassee (1972)

Odum, H. T.: Enviroment, Power, and Society. New York: Wiley-Interscience 1971, 331 pp.

Odum, H. T.: An energy circuit language for ecological and social systems: its physical basis. In: Patten, B. C. (ed.). Systems Analysis and Simulation in Ecology. New York: Academic 1972, Vol. II, pp. 140–212

Oviatt, C. A., Gall, A. L., Nixon, S. W.: Environmental effects of Atlantic menhaden on surrounding water. Ches. Sci. **13**, 321–323 (1972)

Oviatt, C. A., Kremer, P. M.: Predation on the ctenophore, *Mnemiopsis leidyi* by butterfish, *Peprilus triacanthus* in Narragansett Bay, Rhode Island. Ches. Sci. **18**, 236–240 (1977)

Oviatt, C. A., Nixon, S. W.: The demersal fish of Narragansett Bay: an analysis of community structure, distribution, and abundance. Est. Coast. Mar. Sci. **1**, 361–378 (1973)

Oviatt, C. A., Nixon, S. W.: Sediment resuspension and deposition in Narragansett Bay. Est. Coast. Mar. Sci. **3**, 201–217 (1975)

Paasche, E.: Marine algae grown with light-dark cycles. II. *Ditilum brightwellii* and *Nitzschia turgidula*. Physiol. Plantarum **21**, 66–77 (1968)

Paasche, E.: Silicon and the ecology of marine plankton diatoms. I. *Thalassiosira pseudonana* grown in a chemostat with silicate as limiting nutrient. Mar. Biol. **19**, 117–126 (1973)

Paasche, E.: Silicon and the ecology of marine plankton diatoms. II. Silicate uptake kinetics in five diatom species. Mar. Biol. **19**, 262–269 (1973a)

Paffenhöfer, G. A.: Cultivation of *Calanus helgolandicus* under controlled conditions. Helgolander Wiss. Meeres. **20**, 346–359 (1970)

Pandian, T. J.: Intake, digestion, absorption and conversion of food in the fishes *Megalops cyprinoides* and *Ophiocephalus striatus*. Mar. Biol. **1**, 16–32 (1967)

Parsons, T. R., LeBrasseur, R. J.: The availability of foods to different trophic levels in the marine food chain. Contrib. No. 4. Symposium on Mar. Food Chains. Univ. Aarhus, Denmark, 23–26 July, 1968

Parsons, T. R., LeBrasseur, R. J., Fulton, J. D.: Some observations on the dependence of zooplankton grazing on the cell size and concentration of phytoplankton blooms. J. Ocean. Soc. Jap. **23**, 10–17 (1967)

Parsons, T. R., LeBrasseur, R. J., Fulton, J. D., Kennedy, O. D.: Production studies in the Strait of Georgia. Part II. Secondary production under the Fraser River plume, Feb. to May 1967. J. Exp. Mar. Biol. Ecol. **3**, 39–50 (1969)

Parsons, T. R., Stephens, K., Strickland, J. D. H.: On the chemical composition of 11 species of marine phytoplankters. J. Fish. Res. Bd. Can. **18**, 1001–1016 (1961)

Patten, B. C.: Mathematical models of plankton production. Int. Revue ges. Hydrobiol. **53**, 357–408 (1968)

Patten, B. C. (ed.): Systems Analysis and Simulation in Ecology. New York: Academic 1971 1975, Vols. I–III

Patten, B. C.: Coda. In: Patten, B. C. (ed.). System Analysis and Simulation in Ecology. New York: Academic 1971, Vol. I, pp. 583–584

Peters, R., Lean, D.: The characterization of soluble phosphorus released by limnetic zooplankton. Limnol. Oceanogr. **18**, 270–279 (1973)

Petipa, T. S.: The diurnal feeding rhythm of the copepod crustacean *A. clausi*. Dokl. Akad. Nauk USSR **120**, 435–437 (1958)

Petipa, T. S.: Feeding of the copepod *A. clausi*. Transactions of the Sevastopol Biol. Sta. **11**, 72–100 (1959)

Petipa, T. S.: Relationship between growth, energy metabolism, and ration in *A. clausi*. Physiology of Marine Animals, Akad. Nauk, USSR, Oceanographical Commision, 1966, pp. 82–91

Phelps, D. K.: A quantitative study of the infauna of Narragansett Bay in relation to certain physical and chemical aspects of their environments. M. S. Thesis, Univ. R. I., Kingston, Rhode Island (1958)

Pomeroy, L. R., Mathews, H. M., Min, H. S.: Excretion of phosphate and soluble organic phosphorus compounds by zooplankton. Limnol. Oceanogr. **8**, 50–55 (1963)

Pomeroy, L. R., Shenton, L. R., Jones, R. D. H., Reimold, R. J.: Nutrient flux in estuaries. In: Nutrients and eutrophication. ASLO Special Symp., 1972, Vol. I, pp. 274–291

Pratt, D. M.: The phytoplankton of Narragansett Bay. Limnol. Oceanogr. **4**, 425–440 (1959)

Pratt, D. M.: The winter–spring diatom flowering in Narragansett Bay. Limnol. Oceanogr. **10**, 173–184 (1965)

Pritchard, D. W.: Two-dimensional models. In: Estuarine Modeling: An Assessment. Texas: Tracor, Inc., Austin, Report to the Office of Water Qual., E.P.A., 1971

Raymont, J. E. G., Gauld, D. T.: The respiration of some planktonic copepods. J. Mar. Bio. Assoc. U. K. **29**, 681–693 (1951)

Raymont, J. E. G., Miller, R. S.: Production of marine zooplankton with fertilization in an enclosed body of sea water. Int. Rev. Gesam. Hydrobiol. **47**, 169–209 (1962)

Reifsnyder, W. E., Lull, H. W.: Radient energy in relation to forests. Tech. Bull. No. 1344, U. S. Dept. Agr., Washington, D.C. (1965)

R.I. Dept. of Natural Resources, Division of Conservation. Information leaflets 1965–1973. Shellfish Surveys of Narragansett Bay

Rice,T.R., Smith,R.J.: Filtering rates of the hard clam determined with radioactive phytoplankton. Fishery Bull. **58**, 73–82 (1958)

Riley,G.A.: Factors controlling phytoplankton populations on George's Bank. J. Mar. Res. **6**, 54–73 (1946)

Riley,G.A.: A theoretical analysis of the zooplankton population of George's Bank. J. Mar. Res. **6**, 104–113 (1947)

Riley,G.A.: Seasonal fluctuations of the phytoplankton population in New England coastal waters. J. Mar. Res. **6**, 114–125 (1947a)

Riley,G.A.: Oceanography of Long Island Sound 1952–54. II. Physical oceanography. Bull. Bing. Ocean. Coll. **15**, 15–46 (1956)

Riley,G.A.: Theory of food-chain relations in the ocean. In: Hill,M.N. (ed.). The Sea. New York: Wiley-Interscience, 1963, Vol. II, pp. 438–463

Riley,G.A.: The plankton of estuaries. In: Lauff,G.H. (ed.). Estuaries. AAAS pub. No. 83, 1967, pp. 316–326

Riley,G.A.: Transparency-chlorophyll relations. Limnol. Oceanogr. **20**, 150–152 (1975)

Riley,G.A., Bumpus,D.F.: Phytoplankton-zooplankton relationships on George's Bank. J. Mar. Res. **6**, 33–47 (1946)

Riley,G.A., Stommel,H., Bumpus,D.F.: Quantitative ecology of the plankton of the Western North Atlantic. Bull. Bing. Oceanogr. Coll. **12**, 1–169 (1949)

Rosenzweig,M.L.: Paradox of enrichment: destabilization of exploitation ecosystems in ecological time. Science **171**, 385–387 (1971)

Russell,H.J., Jr.: Use of a commercial dredge to estimate a hard shell clam population by stratified random sampling. J. Fish. Res. Bd. Can. **29**, 1731–1735 (1972)

Ryther,J.H.: Photosynthesis in the ocean as a function of light intensity. Limnol. Oceanogr. **1**, 61–70 (1956)

Ryther,J.H., Dunstan,W., Tenore,K., Huguenin,J.: Controlled eutrophication—increasing food production from the sea by recycling human wastes. Biosci. **22**, 144–152 (1972)

Ryther,J.H., Menzel,D.W.: Light adaptation by marine phytoplankton. Limnol. Oceanogr. **4**, 492–497 (1959)

Ryther,J.H., Yentsch,C.S.: The estimation of phytoplankton production in the ocean from chlorophyll and light data. Limnol. Oceanogr. **2**, 281—286 (1957)

Saila,S.B., Flowers,J.M., Cannario,M.T.: Factors affecting the relative abundance of *Mercenaria mercenaria* in the Providence River, R. I. Proc. Natl. Shellfisheries Assoc. **57**, 83–89 (1967)

Sameoto,D.D.: Yearly respiration rate and estimated energy budget for *Sagitta elegans*. J. Fish. Res. Bd. Can. **29**, 987–996 (1972)

Schenk,H., Davis,A.: A turbidity survey of Narragansett Bay. Mimeo Report, Dept. Ocean. Engineering, Univ. R.I., Kingston, Rhode Island (1972)

Schultz,D.M.: Source, formation, and composition of suspended lipoidal material in Narragansett Bay, Rhode Island. Ph. D. Thesis, Univ. R. I., Kingston, Rhode Island (1974)

Scientific Subroutine Package. International Business Machines Programmer's Manual GH 20-0205-4 (1970)

Shaler,N.S., Woodworth,J.B., Foerste,A.F.: Geology of the Narragansett Basin. Washington, D.C.: U.S. Govt. Printing Office, 1899

Short,F.T.: A simulation model of the seagrass production system. In: Studies of the Seagrass ecosystem. Phillips,R.C., McRoy,G.P. (eds.) New York: Marcel Dekker, Inc., in press, Chap. 17

Slobodkin,L.B.: Comments from a biologist to a mathematician. In: Ecosystem Analysis and Prediction. Levin,S.A. (ed.). Proc. SIAM-SIMS Conf., Alta, Utah, 1975, pp. 318–329

Smayda,T.J.: Phytoplankton studies in lower Narragansett Bay. Limnol. Oceanogr. 2, 342–359 (1957)

Smayda,T.J.: Experimental observations on the influence of temperature, light, and salinity on cell division of the marine diatom, *Detonula confervacea* (Cleve) Gran. J. Phycol. **5**, 150–157 (1969)

Smayda, T.J.: The suspension and sinking of phytoplankton in the sea. Oceanogr. Mar. Biol. Ann. Rev. **8**, 353–414 (1970)

Smayda, T.J.: A survey of phytoplankton dynamics in the coastal waters from Cape Hatteras to Nantucket. In: Coastal and Offshore Environmental Inventory, Cape Hatteras to Nantucket Shoals. Kingston: Marine Pub. Ser. No. 2, Univ. R.I., 1973

Smayda, T.J.: The growth of *Skeletonema costatum* during a Winter–Spring bloom in Narragansett Bay, R. I. Norw. J. Bot. **20**, 219–247 (1973a)

Sorokin, Y.I., Panov, D.A.: Balance of consumption and expenditure of food by larvae of Bream at different stages of development. Dokl. Akad. Nauk USSR **165**, 796–799 (1965)

Sorokin, Y.I., Snopkov, V.G., Grimberg, V.M.: The determination of the relation between phytoplankton photosynthesis and submarine illumination in the waters of the central part of the Atlantic. C.R. Acad. Sci. U.S.S.R. **124**, 432–435 (1959)

Steele, J.H.: Plant production in the northern North Sea. Scottish Home Dept. Mar. Res., No. **7**, 1–36 (1958)

Steele, J.H.: Environmental control of photosynthesis in the sea. Limnol. Oceanogr. **7**, 137–150 (1962)

Steele, J.H.: Notes on some theoretical problems in production ecology. In: Primary Productivity in Aquatic Environments. Goldman, C.R. (ed.). Mem. Ist. Ital. Idrobiol., 18 Suppl., Berkeley: Univ. Calif. 1965, pp. 383–398

Steele, J.H.: The Structure of Marine Ecosystems. Cambridge, Massachusetts: Harvard Univ., 1974

Steele, J.H., Baird, I.R.: Relations between primary production, chlorophyll and particulate carbon. Limnol. Oceanogr. **6**, 68—78 (1961)

Steeman-Nielsen, E., Hansen, V.K., Jorgensen, E.G.: The adaptation to different light intensities in *Chlorella vulgaris* and the time dependence on transfer to a new intensity. Physiol. Plantarum **15**, 505–517 (1962)

Steeman-Nielsen, E., Jensen, A.: Primary oceanic production. Galathea Rep. **1**, 49–136 (1957)

Steeman-Nielsen, E., Park, T.S.: On the time course of adapting to low light intensities in marine phytoplankton. J. Cons. Int. Explor. Mer **29**, 19–24 (1964)

Stickney, A.P., Stringer, L.D.: A study of the invertebrate fauna of Greenwich Bay, R.I. Ecology **38**, 111–122 (1957)

Strickland, J.D.H.: Measuring the production of marine phytoplankton. II. Chemical composition of phytoplankton. Fish. Res. Bd. Can. Bull. no. 122 (1966)

Sushchenya, L.M.: Food rations, metabolism, and growth of crustaceans. In: Steele, J.H. (ed.). Marine Food Chains. Berkeley: Univ. Calif., 1970

Sverdrup, H.U., Johnson, M.W., Fleming, R.H.: The Oceans. Englewood Cliffs, N.J.: Prentice Hall, 1947

Talling, J.F.: The underwater light climate as a controlling factor in the production ecology of freshwater phytoplankton. Mitt. Internat. Verein. Limnol. Oceanogr. **19**, 214–243 (1971)

Thomas, W.H., Dodson, A.N.: Effects of phosphate concentration on cell division and yield of a tropical oceanic diatom. Biol. Bull. **134**, 199–208 (1968)

Ulanowictz, R.E., Flemer, D.A., Heinle, D.R., Mobley, C.D.: The *a posteriori* aspects of estuarine modeling. In: Cronin, J. (ed.) Estuarine Research. Proc. 2nd Int. Est. Res. Conf., Myrtle Beach, S.C. New York: Academic 1975, Vol. I, pp. 601–616

U.S. Army Corps of Engineers: Contamination dispersion in estuaries, Narragansett Bay. Hydraulic Model Investigation. U.S. Army Eng. Waterways Expt. Sta. Misc. Paper 2–332, Report 2. Vicksburg, Mississippi (1959)

U.S. Geological Survey. Surface Water Records for Mass., New Hampshire, Rhode Island and Vermont. Annual.

U.S. Weather Bureau. Local Climatological Data, Providence. R.I. Yearly Summaries.

Vargo, G.A.: The influence of grazing and nutrient excretion by zooplankton on the growth and production of the marine diatom, *Skeletonema costatum* (Greville) Cleve in Narragansett Bay. Ph. D. Thesis, Univ. R. I., Kingston, Rhode Island (1976)

Verduin, J.: Hard clam pumping rates: energy requirement. Science **166**, 1309–1310 (1969)

Vinogradov, A.P.: The elementary chemical composition of marine organisms. Memoir II, Sears Foundation for Marine Research, New Haven, Connecticutt. p. 647 (1953)

Vollenweider, R. A.: Calculation models of photosynthesis depth curves and some implications regarding day rate estimates in primary production measurements. In: Goldman, C. R. (ed.) Primary Productivity in Aquatic Environments. Mem. Ist. Ital. Idrobiol., 18 Suppl., Berkely: Univ. Calif., 1965, pp. 427–457

Volterra, V.: Variations and fluctuations of the number of individuals in animal species living together. In: Chapman, R.N. (ed.). Animal Ecology. New York: McGraw-Hill 1926, pp. 409–448

Waite, T., Mitchell, R.: The effect of nutrient fertilization on the benthic alga *Ulva lactuca*. Botanica Marina **15**, 151–156 (1972)

Walne, P. R.: The influence of current speed, body size, and water temperature on the filtration rate of five species of bivalves. J. Mar. Biol. Assoc. U.K. **52**, 345–374 (1972)

Walsh, J. J.: A spatial simulation model of the Peruvian upwelling ecosystem. Deep-Sea Res. **22**, 201–236 (1975)

Walsh, J. J., Dugdale, R. E.: A simulation model of the nitrogen flow in the Peruvian upwelling system. Invest. Pesq. **35**, 309–330 (1971)

Weisberg, R. H., Sturges, W. III.: The net circulation in the West Passage of Narragansett Bay. Kingston: Tech. Rept. No. 3–73, Univ. R.I., 1973, 93 pp.

Wiegert, R. G.: Simulation models of ecosystems. Ann. Rev. Ecol. Syst. pp. 311–338 (1975)

Williams, P.J. LeB.: The validity of the application of simple kinetic analysis to heterogeneous microbial populations. Limnol. Oceanogr. **18**, 159–164 (1973)

Winberg, G. G.: Methods for the Estimation of Production in Aquatic Animals. London: Academic. Trans. from Russian, Duncan, A., 1971

Yentsch, C. S.: Critical mixing depth—a problem in the measurement of respiration, In: Cech, J. J. Jr., Bridges, D. W., Horton, D. B. (eds.). Respiration of Marine Organisms. South Portland, Maine: TRIGOM 1975, pp. 1–10

Yentsch, C. S., Lee, R. W.: A study of photosynthetic light reactions, and a new interpretation of sun and shade phytoplankton. J. Mar. Res. **24**, 319–337 (1966)

Zeuthen, E.: Oxygen uptake and body size in organisms. Quart. Rev. Biol. **28**, 1–12 (1953)

Zillioux, E.: Ingestion and assimilation in laboratory cultures of *Acartia*. Tech. Rept., Nat. Mar. Water Qual. Lab., EPA, Narragansett, R.I. (1970)

Author and Subject Index

Vol. 15: **Plants in Saline Environments**
Editors: A. Poljakoff-Mayber, J. Gale
54 figs. VII, 213 pages. 1975

Vol. 16: **Fennoscandian Tundra Ecosystems**
Part 1: *Plants and Microorganisms*
Editor: F. E. Wielgolaski
Editorial Board: P. Kallio, T. Rosswall
90 figs., 96 tables. XV, 366 pages. 1975

Vol. 17: **Fennoscandian Tundra Ecosystems**
Part 2: *Animals and Systems Analysis*
Editor: F. E. Wielgolaski
Editorial Board: P. Kallio, H. Kauri, E. Østbye, T. Rosswall
81 figs., 97 tables. XIII, 337 pages. 1975

Vol. 18: **Remote Sensing for Environmental Sciences**
Editor: E. Schanda
178 figs. (7 color-plates), 31 tables. XIII, 367 pages. 1976

Vol. 19: **Water and Plant Life**
Problems and Modern Approaches
Editors: O. L. Lange, L. Kappen, E.-D. Schulze
178 figs., 66 tables. XX, 536 pages. 1976

Vol. 20: F. B. Christiansen, T. M. Fenchel
Theories of Populations in Biological Communities
68 figs., 5 tables. X, 144 pages. 1977

Vol. 21: V. Ya. Alexandrov
Cells, Molecules and Temperature
Conformational Flexibility of Macromolecules and Ecological Adaptation
Translated from the Russian by V. A. Bernstam
74 figs., 25 tables. XI, 330 pages. 1977

Vol. 22: R. Guderian
Air Pollution
Phytotoxicity of Acidic Gases and Its Significance in Air Pollution Control. Translated from the German by C. J. Brandt
40 figs., 4 in color, 26 tables. VIII, 127 pages. 1977

Vol. 23: F. D. Por
Lessepsian Migration
The Influx of Red Sea Biota into the Mediterranean by Way of the Suez Canal
47 figs., 2 maps, 10 plates. Approx. 250 pages. 1977

Scheduled to come out in December 1977

Vol. 25: **Microbial Ecology of a Brackish Water Environment**
Editor: G. Rheinheimer
Approx. 77 figs., 84 tables. Approx. 300 pages

This book describes the results of a 15-month investigation in the Kiel Bight (Baltic Sea) in which more than 50 hydrographical, chemical, and microbiological parameters were measured. Using methods such as, for example, scanning electron microscopy and autoradiography, a comparative analysis of most of the measured data on the bacterial colonization of detritus and the relations between pollution, production, and remineralization led to the energy fluxes. Together with Vol. 24 by Kremer and Nixon about the ecological studies at the Narragansett Bay and the system-simulation by a computer model, these two books represent a comprehensive study of coastal marine ecosystems.

Springer-Verlag
Berlin
Heidelberg
New York

45942233R00132

Made in the USA
Lexington, KY
16 October 2015